Unstructured Cellular Automata in Ecohydraulics Modelling

非结构元胞自动机

生态水力学模拟

Unstructured Cellular Automata
in Ecohydraulics Modelling

DISSERTATION

Submitted in fulfillment of the requirements of
the Board for Doctorates of Delft University of Technology
and
of the Academic Board of the UNESCO-IHE
Institute for Water Education

for
the Degree of DOCTOR
to be defended in public on
Tuesday, 7 October 2014, at 15:00 hours
in Delft, the Netherlands

by

Yuqing LIN

Master of Science in Water Science and Engineering (with distinction)
UNESCO-IHE, Delft, the Netherlands

born in Yangzhou, China

This dissertation has been approved by the supervisors:
Prof. dr. ir. A.E. Mynett
Prof. dr. Q.W. Chen

Composition of Doctoral Committee:

Chairman	Rector Magnificus TU Delft
Vice-Chairman	Rector UNESCO-IHE
Prof. dr. ir. A.E. Mynett	UNESCO-IHE / Delft University of Technology, supervisor
Prof. dr. Q.W. Chen	Nanjing Hydraulic Research Institute, China, supervisor
Prof. dr. ir. G.S. Stelling	Delft University of Technology
Prof. dr. M.E. McClain	UNESCO-IHE / Delft University of Technology
Prof. dr. ir. H.X. Lin	Delft University of Technology / Tilburg University
Dr. H. Li	Int. Water Association, the Netherlands
Prof. dr. ir. M.J.F. Stive	Delft University of Technology, reserve member

CRC Press/Balkema is an imprint of the Taylor & Francis Group, an informa business

Published by:
CRC Press/Balkema
PO Box 11320, 2301 EH Leiden, The Netherlands
e-mail: Pub.NL@taylorandfrancis.com
www.crcpress.com – www.taylorandfrancis.com

ISBN 978-1-138-02740-4 (Taylor & Francis Group)

To my grandmother Xiaomei Liu and my Parents
without their support over the years none of this would be possible

Summary

The field of ecohydraulics has emerged over recent decades and addresses the interactions between *hydro*-dynamic and *eco*-dynamic processes. While hydrodynamic processes are usually well described by mathematical equations based on physical principles of mass, momentum and energy conservation at the *global* system level, ecosystem dynamics often involve specific *local* interactions between species with a strong dependence on individual properties. Thus, while partial differential equations may be well suited to describe the hydraulic (hydrodynamic) features, a more localized mathematical description is often preferred when modelling the ecosystem behaviour. It has been observed in the field that even in case of relatively simple interactions at the local scale, ecological evolution often leads to quite complex spatio-temporal patterns at the global scale.

Cellular Automata (CA) are known to be capable of representing (strongly) non-linear dynamical systems at discrete levels in time and space. Von Neumann (1949) was first to develop a numerical algorithm for digital computers considering an array of cells in a lattice arrangement. He observed that repetitive use of simple local transition rules could evolve into quite complex space-time patterns. The concept of CA was further elaborated by (Wolfram, 1984) and applications were investigated in a wide range of fields in science and engineering, ranging from turbulence theory to oil exploration, and from information theory to population dynamics. Applications in the field of ecohydraulics were explored by (Minns et al., 2000) who showed CA to be a viable paradigm for ecosystem modelling due to its ease of accounting for local differences in individual species properties, while at the same time enabling complex space-time dynamics to emerge at the global scale.

Cellular automata constitute a mathematical system in which each cell starts from an initial state and all cells update their states synchronously at discrete steps according to simple local rules. The classical configuration of cellular automata consists of an array of identical and uniformly distributed cells in a structured grid arrangement. But in the field of hydrodynamics, the use of unstructured grids has become more and more popular due to its capability to deal with arbitrary geometries and its flexibility to adapt to changing boundary configurations.

In this thesis, the concept of Unstructured Cellular Automata (UCA) is explored for unstructured computational meshes of varying sizes and topological arrangements. The main objective of the research is to identify whether the cellular automata modelling paradigm can be applied to unstructured computational grid arrangements as well. Various options of unstructured cellular automata configurations are explored and their performance is investigated by numerical experiments. The practical applicability of UCA in eco-hydraulics modelling is explored through a number of case studies and compared with field measurements.

Compared with classical cellular automata, UCA configurations often contain different cell-types within the same computational grid: the computational stencil for any particular object cell may differ from the ones of its neighbours. The properties of such UCAs are analysed in some detail in this thesis. First, the characteristics of two types of cell configurations (triangular elements and polygon elements) were explored via numerical experiments. Then, the effect of initial conditions was analysed for various initial spatial distributions and different initial percentages in a prey-predator type application. Furthermore, the performance of different Neighbourhood Schemes for UCA was investigated, exploring the analogies with the conventional Neumann, Moore and Extended Moore configurations in classical CA.

Cellular Automata are sometimes considered as an alternative approach in numerical modelling. There are many references to, for example, cellular automata being used to solve the Navier-Stokes equation by Lattice-gas CA. However, Lattice-gas CA holds a microscopic view and is thus restricted to very small scales. For real applications, Lattice-gas CA is less practical for solving the full Navier-Stokes equations (Chen, 2004). In this thesis special attention was paid to exploring the differences and analogies between CA and PDEs.

Partial Differential Equations (PDEs) can have significant advantages like the ability to obtain analytical solutions that capture the main features of the problem at hand in continuous time and space coordinates. But when solutions of PDEs are obtained by digital computer based on discrete approximations, the solution also becomes discrete in space-time. Cellular Automata on the other hand are already fully discrete by the nature of its concept as described above. The analogies between discrete solutions of PDEs and equivalent approaches from CA are discussed in this thesis. Particular attention is given on how to deduce transition rules for CA-modelling from PDE-based approaches using finite difference methods. Some specific types of partial differential equations (diffusion equation, wave equation, Burgers equation) are used as a reference for evaluating the equivalent performance of CA simulations.

Since UCA can have varying neighbourhood properties in contrast with classical CA, the influence of cell size in UCA was analysed in this thesis (chapter 3) using the Finite Volume Method. A characteristic parameter —min distance of UCA– was put forward and tested by several numerical experiments on different types of meshes (rough meshes, locally refined meshes, globally refined meshes) in order to explore the usefulness of this parameter.

The analysis results from computational theory of UCA were applied to a water quality modelling case study of spiked pollution loading for Hong Kong Bay (Chapter 5). In Chapter 6 UCA was used to quantify the spatial distribution of macro invertebrates as part of a river restoration study in China, simulating the spatial evolution of benthonic macro invertebrates under different flow regulation scenarios. In Chapter 7 UCA is applied to an aquatic pond system and compared with Individual-based modelling. The main findings of this research are summarised in Chapter 8.

Samenvatting

Het vakgebied van de ecohydraulica is gedurende de afgelopen decennia ontstaan en richt zich op de interakties tussen *hydro*-dynamische en *eco*-dynamische processen. Waar de hydrodynamische processen gewoonlijk goed kunnen worden beschreven door wiskundige vergelijkingen gebaseerd op fysische behoudswetten voor massa, impuls en energie voor het *totale* systeem, gaat het bij de dynamica van ecosystemen veel meer om *lokale* interacties tussen verschillende soorten die sterk afhankelijk zijn van hun specifieke eigenschappen.

Vandaar dat partiële differentiaalvergelijkingen uitstekend geschikt zijn om hydraulische (hydrodynamische) verschijnselen te beschrijven, maar een meer lokale wiskundige benadering de voorkeur verdient bij het modelleren van ecosystemen. Waarnemingen in de natuur geven aan dat zelfs betrekkelijk eenvoudige interacties op lokale schaal kunnen leiden tot complexe patronen op grotere schaal. Cellulaire Automata (CA) staan bekend om hun vermogen om complexe niet-lineaire dynamische systemen te kunnen weergeven op discrete niveaus in tijd en ruimte. Von Neumann (1949) ontwikkelde als eerste een numeriek algoritme voor een digitale computer op een gestructureerd rekenrooster. Het viel hem op dat het herhaald toepassen van relatief eenvoudige lokale regels tot complexe ruimte-tijd patronen kon leiden. Het concept van cellulaire automata werd verder onderzocht door (Wolfram, 1984) met een breed scala aan toepassingen in wetenschap en praktijk, van turbulentie theorie tot oliewinning, en van informatie theorie tot populatie dynamica.

Toepassingen op het gebied van ecohydraulica werden onderzocht door (Minns et al., 2000) die aantoonden dat CA mogelijkheden bood voor het modelleren van ecosystemen aangezien het gemakkelijk rekening kon houden met lokale verschillen in eigenschappen van soorten, terwijl het tegelijkertijd het gedrag van complexe dynamische systemen kon simuleren op grotere ruimte en tijdschalen. Cellulaire Automata zijn wiskundige systemen waarbij iedere cel start met een bepaalde begincondite en alle cellen hun toestand tegelijkertijd opwaarderen op basis van lokale regels (gerelateerd aan de toestand van naburige cellen). Een klassieke CA configuratie bestaat uit een regelmatig rooster van identieke cellen die uniform verdeeld zijn. Maar in de hydrodynamica neemt het gebruik van ongestructureerde rekenroosters snel toe, vanwege de mogelijkheden om willekeurige geometrieën te kunnen representeren en zich te kunnen aanpassen aan veranderende randvoorwaarden.

In dit proefschrift wordt onderzoek gedaan naar de mogelijkheid om het concept van cellulaire automata toe te passen op ongestructureerde rekenroosters. Verschillende vormen van ongestructureerde rekenroosters worden onderzocht en hun toepasbaarheid nagegaan middels numerieke simulaties. De praktische toepasbaarheid op het gebied van ecohydraulische modellen wordt nagegaan aan de hand van metingen in praktijksituaties.

Vergeleken met de klassieke CA structuren kunnen ongestructureerde cellulaire automata verschillende celconfiguraties hebben binnen hetzelfde rekengrid: de rekenstencils kunnen per buurcel verschillen. De eigenschappen van verschillende ongestructureerde configuraties worden in dit proefschrift onderzocht. Allereerst wordt het verschil tussen driehoekige cellen en polygon elementen nagegaan aan de hand van numerieke simulaties. Vervolgens wordt de invloed van de begincondities onderzocht voor verschillende ruimtelijke verdelingen en verschillende beginpercentages in een jager-prooi configuratie. Vervolgens worden de analogieën onderzocht met de bekende Neumann, Moore en Extended Moore configuraties uit de klassieke CA.

CA wordt soms gezien als een alternatieve vorm van numeriek modelleren. Zo zijn er bijv. meerdere toepassingen bekend van het oplossen van de Navier-Stokes vergelijkingen middels Lattice-gas CA. Deze zijn echter vaak beperkt tot gebieden met kleine afmetingen. Voor praktische toepassingen biedt CA minder mogelijkheden (Chen, 2004). In dit proefschrift wordt speciaal aandacht besteed aan het nagaan van de overeenkomsten en verschillen tussen cellulaire automata en numerieke oplosmethoden voor partiële differentiaalvergelijkingen. PDEs hebben duidelijke voordelen waaronder het hebben van analytische oplossingen die de belangrijkste kenmerken in zich dragen. Maar als numerieke technieken worden gebruikt op digitale computers, dan worden de oplossingen discreet benaderd in ruimte en tijd. Cellulaire Automata daarentegen zijn uit zichzelf al volledig discreet. De overeenkomsten en verschillen worden in dit proefschrift nagegaan.

Daarbij wordt in het bijzonder aandacht besteed aan het afleiden van lokale CA-regels vanuit het gedrag van PDE oplossingen op basis van eindige differentieschema's. Een aantal voorbeelden (diffusievergelijking, golfvergelijking, Burgers' vergelijking) wordt gebruikt als referentie voor het equivalente gedrag van cellulaire automata. Aangezien ongestructureerde CA lokaal verschillende stencils kan hebben (in tegenstelling tot gestructureerde CA), wordt het effect van verschillende stencils nagegaan in analogie met de Eindige Volume Methode (hoofdstuk 3). Voorgesteld wordt om een karakteristieke parameter –minimale afstand– te gebruiken en de relevantie daarvan wordt in dit proefschrift aan de hand van numerieke experimenten onderzocht.

De toepasbaarheid van de hier ontwikkelde theorie is nagegaan aan de hand van een waterkwaliteitsstudie ten gevolge van afvalwaterlozing in de baai van Hong Kong (hoofdstuk 5). In hoofdstuk 6 wordt de ruimtelijke verdeling van ongewervelde soorten onderzocht bij verschillende scenario's van rivierbeheer in China. Hoofdstuk 7 laat aan de hand van een toepassing in een kleine vijver zien dat ongestructureerde cellulaire automata veel overeenkomsten hebben met individual-based modelling (IBM). De belangrijkste bevindingen van dit onderzoek zijn samengevat in Hoofdstuk 8.

Acknowledgements

My sincere appreciation and gratitude goes to all individuals and institutions that supported me during my PhD studies, especially to my sponsors Delft Hydraulics (now part of Deltares) and UNESCO-IHE for their financial support that made my study become reality.

My deepest gratitude goes first and foremost to my promoter Prof. Arthur Mynett, who has initially proposed my PhD study and has guided me throughout my studies. His experience, warmness and his sense of responsibility were essential in all stages of my study. Both his broad directional advice and detailed technical instructions have given me the strength and endurance to pursue my study. In particular during the time when I suffered from health problems and felt rather helpless, he secured medical and funding support for me. I am deeply grateful for all his guidance and support.

I would also like to give special thanks and appreciation to my co-promoter Prof. Qiuwen Chen, both during my time at the Chinese Academy of Science Research Centre for Eco-Environmental Science (CAS-RCEES) and later at the Nanjing Hydraulics Research Institute (NHRI), for his continuous guidance, systematic supervision, sharing his vision and wisdom with me, trusting my capabilities and giving me freedom to develop new ideas. Without his illuminating instructions and valuable ideas, this study would not have been finished successfully.

Sincere thanks should also be given to Prof. Guus Stelling of TUDelft, for all his constructive suggestions and support. The discussions with him are very much appreciated. Special thanks as well go to Ir. Leo Postma and other experts of Deltares, for giving me very helpful suggestions, and directions, and for sharing their valuable practical experience with me.

I am deeply thankful to Dr. Hong Li, Dr. Qinghua Ye, Dr. Ken Wong and Dr. Li Wang for generously sharing with me their reference articles and reports, and for their helpful suggestions and good advisory comments. Also I would express my thanks to so many of my friends: Taopin Wan, Shengyang Li, Yuanyang Wan, Liqin Zuo, Leicheng Guo, Chunqing Wang, Xuan Zhu, Hui Chen, Hui Qi, Zhuo Xu, Kun Yan, Lihe Yin, Xueqing Cao, Xiaofeng Luo, Yufang Han and many more, who shared the good time with me when I was in Delft.

I would also send my special thanks to my friends in Beijing and Nanjing: Ruonan Li, Zhijie Li, Rui Han, Xiaoqing Zhang, Ruihua Liu and Xiaofeng Liu, who always cared for me a lot. And I should also mention my other Chinese teachers who spent a lot of time in helping me during my studies: Prof. Guangchi Li, Prof. Wen Wang, Prof. Yiqing Guan, Ms. Ning Su, Dr. Liuming Hu, Dr. Weifeng Li, Dr. Guoxian Huang, Dr. Jingfeng Ma. Many thanks to my Chinese colleagues: Qingrui Yang, Qiang Xu, Wei Huang, Fei Ye and my other friends from all over the world, as we help and learn from each other.

I appreciate very much UNESCO-IHE and its staff for helping me in many ways during my studies and my stay in the Netherlands. I also have to acknowledge Hohai University for providing the opportunity for me to study in the international master degree program in hydroinformatics, which undoubtedly extended the scope of my world. I am also grateful to Nanjing Hydraulics Research Institute and the Research

Centre for Eco-Environmental Science of the Chinese Academy of Science in Beijing; they provided me with nice research environments in China, and opportunities to cooperate with my Chinese colleagues, which gave me the chance to improve my academic ability.

Finally, I would like to give my special thanks and express my heartfelt appreciation to my families for their unconditional love and full support: my grandmother Xiaomei Liu, my parents and my husband. Their love always accompanied me and they provided great spiritual power in my heart.

Yuqing Lin
UNESCO-IHE, Delft, the Netherlands
September 2014

Contents

Chapter 1

Introduction

1.1 Background

Cellular Automata are able to represent non-linear dynamical systems at discrete intervals in time and space. First proposed by Von Neumann (1949), the theory and application of CA developed rather slowly. Codd and Langton (1968) proved that CA has self-reproduction capabilities, but CA probably became most well known owing to the famous 'Game of Life' as developed by Conway (1970). After that, Wolfram (1986, 1994) contributed significantly to the further evolution of CA.

The classical concept of Cellular Automata consists of an n-dimensional array of cells. Each cell starts from an initial state and all cell states are updated synchronously at discrete steps according to a simple local rule (Beigy & Meybodi, 2004). The new state of each cell depends on the previous state of the cell itself, and of the states of its nearest neighbours only (James & Kingsbery, 2006). Even when based on a set of simple *local* rules, cellular automata can generate very complex global patterns that are often not unlike complex processes observed in nature.

One could argue that the concept of Cellular Automata is an interdisciplinary modelling paradigm that integrates computability theory, mathematics, theoretical biology and microstructure modelling. It belongs to the computational tools of physicists, mathematicians, computer scientists and biologists alike. Hence there are different interpretations and meanings of cellular automata. From the view of physicists, cellular automata are defined as discrete, infinite-dimensional dynamical systems; mathematicians describe it as a discrete space-time mathematical concept; computer scientists think of it as an emerging artificial intelligence technique while biologists treat cellular automata as an abstract representation of life.

Cellular Automata can be used to study many phenomena, including communications theory, information transfer, computing algorithms, reproduction, competition and evolution (Smith, 1969; Perrier, 1996). For processes related to the theory of dynamical systems such as turbulence, chaos theory, fractal dynamics and other complex dynamic systems, the study of CA provides an effective modelling tool (Vichhac, 1984; Bennett, 1985).

1.2 Research scope

Cellular Automata have become more and more popular due to their conceptual simplicity, ease of computer implementation, and ability to exhibit a variety of complex global spatial pattern evolution dynamics, all stemming from using a set of simple local rules. These features have attracted researchers' attention from a wide range of different fields of science as elaborated in Chapter 2 hereafter.

Applications of CA have been successfully used to model tumor growth, fluid flow, galaxy formation, biological pattern formation, avalanches, traffic jams, parallel computers, image processing, and earthquakes (Hu & Ruan, 2003); urban land-use patterns (Engelen et al., 1993); spreading of forest fires (Karafyllidis & Thanailakis, 1996); random number generation (Seth et al., 2008).

Ecological evolution can generate complex space-time dynamics. Cellular Automata proved to be a viable approach for ecosystem modelling due to its relative ease to implement differences between individual properties and account for local interactions. Examples include the spreading of water plant species (Babovic & Baretta, 1996); the evolution of vegetation (Balzter et al., 1997); animal population dynamics (Minns et al., 2000; Chen et al., 2002b); the spreading of marine macrophytes (Wortmann et al., 1997; Mynett & Chen, 2004), prey-predator population dynamics (Chen & Mynett, 2003) and harvesting strategies (Chen, 2006).

CA based ecological models usually define simple interactions between components at a local level that can lead to complex spatial patterns emerging on a global scale. (Wootton, 2001; Chen & Mynett, 2003). These features of CA caused increasing popularity in ecohydraulics modelling of coupled hydro-dynamic and eco-dynamic processes.

1.3 Objectives and research questions

The classical concept of CA is based on structured computational grids. In recent years, however, unstructured grids have become more and more popular in hydrodynamic computations due to its flexibility to handle irregular geometries and capability to create local grid refinement where needed. In ecohydraulics modelling the computational meshes are usually first generated for the hydrodynamic flow and transport processes, and then used for ecosystem modelling.

In order to avoid mapping results from unstructured grid onto structured grids in order to be able to use cellular automata computing, an Unstructured Cellular Automata (UCA) approach is explored in this thesis.

While the classical CA concept consists of an array of identical interacting cells, the UCA concept as developed in this thesis consists of object cells and its neighbours that differ from each other. (Lin et al., 2008)

The main objective of this research is to explore an unstructured cellular automata modelling approach and to investigate the performance of unstructured cellular automata by numerical experiments and explore possible applications of unstructured cellular automata in eco-hydraulics modelling.

More specifically, the objectives of this research are:
1. To explore unstructured cellular automata (UCA) schemes;
2. To evaluate the performance of UCA using numerical experiments;
3. To investigate similarities / differences between numerical discretisation of Partial Differential Equations and UCA modelling;
4. To explore the computational theory of unstructured cellular automata;
5. To apply the unstructured cellular automata paradigm to ecological modelling;
6. To compare unstructured cellular automata with other computational methods.

1.4 Thesis Outline

The thesis is structured in 8 chapters where

Chapter 1 introduces the research background and motivation, the description of the research problems and the related methodologies.

Chapter 2 depicts the fundamental aspects of the classical cellular automata modelling paradigm.

Chapter 3 describes the concepts of Unstructured Cellular automata with construction method and sensitivity analysis.

Chapter 4 elaborates computational theory of unstructured cellular automata including the relationship between UCA and PDE, and deriving rules for cellular automata from the analogous differential equations.

Chapter 5 presents several applications of unstructured cellular automata to ecosystem including spatial water quality modelling for spiked pollution loading.

Chapter 6 demonstrates the procedure of how to use unstructured cellular automata to quantify the evolution of benthonic macroinvertebrate under flow regulation.

Chapter 7 compares the concepts of Individual-Based Modelling (IBM) and spatially-based unstructured cellular automata in aquatic ecosystem modelling.

Chapter 8 concludes the research activities and highlights the findings of the research, In addition, unsolved problems and recommendations for future exploration are outlined.

Chapter 2

The concept of Cellular Automata

2.1 A brief historical overview

Von Neumann (1949) first introduced the concept of Cellular Automata (CA) for digital computers using regular lattice architecture of cells. He explored options for self-reproduction considering 29 possible cell states. Since then, the trend has been to focus on investigating the simplest possible configuration which is still capable of (re)producing a complex dynamic system. CA constitute a mathematical system in which many simple components act together at discrete intervals in time and the states of all cells change in parallel. CA often exhibits 'self-organization' behaviour. Even starting from complete disorder, irreversible evolutions can spontaneously generate an ordered structure – and vice versa. Following up from the initial CA concept, applications tend to use a regular grid as the underlying network structure. However, other regular grid configurations could be hexagonal or triangular structures (Navid and Aghababa, 2013)

The classical cellular automata components update their states at discrete time steps according to local evolution rules, which are functions of the states of a cell itself and its immediate neighbours. Fig. 2-1 (a) and (b) illustrate a one-dimensional and a two-dimensional CA configuration with their nearest neighbours. The corresponding evolution rules are expressed in Equations (2-1) and (2-2) resp.

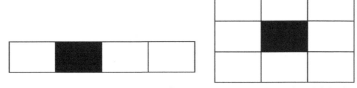

Fig. 2-1 (a) One-dimensional CA   **(b)** Two-dimensional CA.

$$a_i^{t+1} = \phi\left(a_{i-1}^t, a_i^t, a_{i+1}^t\right)$$

$$(2\text{-}1)$$

$$a_{i,j}^{t+1} = \phi\left(a_{i,j}^t, a_{i-1,j-1}^t, a_{i-1,j}^t, a_{i-1,j+1}^t, a_{i,j-1}^t, a_{i,j+1}^t, a_{i+1,j-1}^t, a_{i+1,j}^t, a_{i+1,j+1}^t\right)$$

$$(2\text{-}2)$$

where φ is a function of local evolution rules.

According to the definition of cellular automata, classical cellular automata have the following characteristics (Rothman & Zaleski, 1997; Li, 1997; Chen, 2004):

(1) Homogeneity: all cells follow the same evolution rules;
(2) Discrete in space: Cellular Automata are distributed discrete in space;
(3) Discrete in time: evolution of the CA system is based on step-wise intervals in time;
(4) Parallelism: all cell states are updated simultaneously, which is particularly suitable for parallel computing;
(5) Locality: each cell can only gather information from its neighbours and can only affect its neighbours.

In practical applications many cellular automata models have certain extensions from the basic features. For example for continuous cellular automata all cells have continuous states during the evolution process. But quite often, as pointed out in the conceptual analysis of cellular automata, the homogeneity, parallelism, and locality are the core of the algorithm; any variation on cellular automata usually follows these core features, especially the locality feature.

2.2 Cellular Automata with different structured lattice configurations

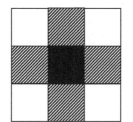

Fig. 2-2 Structured square grids

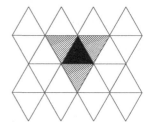

Fig. 2-3 Structured triangle grid

Cellular Automata come in a variety of shapes. The simplest computational 'grid' is a one-dimensional array. Two-dimensional cellular configurations can consist of square, triangular or hexagonal cells. Classical cellular automata are based on structured grids, most commonly square girds. But there are other type of grids such as triangular grids and hexagon automata that also have isotropy.

In the following section, classical cellular automata models are considered based on square grids (Fig.2-2) and structured triangle grids (Fig. 2-3). In order to provide two sets of lattices with the same initial conditions, the structured triangle grids are generated from the square grid (every square grid is divided into two sub-triangular grids). The initial conditions are randomly generated. In order to assign the same initial pattern to triangular and square grids, three kinds of initial conditions are tested, as listed below:

➢ In the first scenario the initial state is set for the square grid, and then the two sub-triangles are assigned the same initial state. As shown in Fig. 2-4, the population dynamics are both stable (the two sub-figures on the right side); and the spatial patterns (the two sub-figures on the left side) are similar.

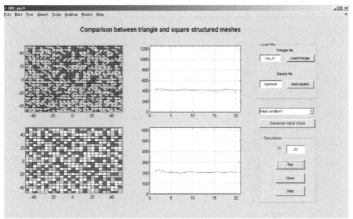

Fig. 2-4 Comparison between triangular and squared structured meshes (Initial condition 1)

➢ In the second scenario random initial values are generated for the triangular grid, then averaged sub-triangles' values are used as input for the square grid.

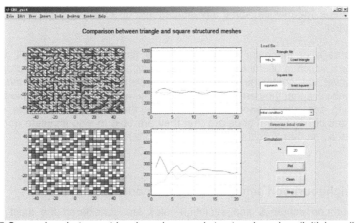

Fig. 2-5 Comparison between triangle and squared structured meshes (Initial condition 2)

In Fig. 2-5 for the square grid part, the influence of initial conditions vanishes rapidly within a few running steps. The two types of grids reach stable states, but in this case the spatial patterns appear to be different.

➢ In the third scenario random initial values are generated for the triangular grid and the rule between every two sub-triangles is executed for a single step (like a local rule for initial distribution). The output is assigned to the square element.

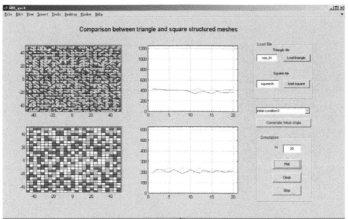

Fig. 2-6 Comparison between triangle and squared structured meshes (Initial condition 3)

In summary, classical cellular automata with different structured lattices are implemented. Generally speaking, there is no obvious difference between the structured triangle grids and the square grids in a regular area, in the sense that both result in stable states. The influence of initial conditions vanishes rapidly within a few running steps. Meanwhile, some advantages and disadvantages of different kinds of grids appear.

An advantage of triangular grids is that they have a relatively small number of neighbours, which is useful in some cases; drawback is that computer programming is not convenient; sometimes one needs to convert to a square grid. Advantages of square grids are that they are intuitive and simple, and particularly suitable for display on existing computers; shortcomings are that they can prefer regular geometry. Advantages of hexagonal grids are that they are able to also simulate non-isotropic phenomena and therefore the model can be more natural and real.

2.3   Neighbourhood schemes of cellular automata

Neighbourhood schemes and evolution rules are considered the most important properties that are worthy of further elaboration (Chen, 2008). The neighbourhood can be defined as a spatial region specified for a cell to gather information from its vicinity when the cell is updating its state (Chen, 2008). There are several neighbourhood schemes belonging to 2-dimension cellular automata, e.g. Von Neumann scheme; Moore scheme; and Margous scheme. Fig. 2-7 shows the most frequently used neighbourhood schemes, named after John von Neumann, who used it in his studies of self-reproducing machines.

(1) Von Neuman Neighborhood schemes

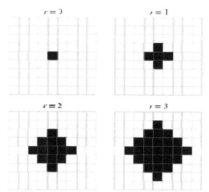

Fig. 2-7 Von Neumann Neighbourhood schemes

A diamond-shaped neighbourhood that can be used to define a set of cells surrounding a given cell (x_0, y_0) that may affect the evolution of a two-dimensional cellular automaton on a square grid, the Von Neumann neighbourhood of range r is defined by

$$N^M_{(x_0,y_0)} = \{(x, y): |x-x_0|+|y-y_0| <= r\}$$

Von Neumann neighbourhoods for ranges r=0, 1, 2, and 3 are illustrated above. The number of cells in the von Neumann neighbourhood of range r is the centred square number $2r(r+1) +1$, the first few of which are 1, 5, 13, 25, 41, 61, ... etc. When the range of the neighbourhood r equals 1 (r=1), especially for the square meshes in d-dimension, a cell has 2^d number of neighbours.

(2) Moore Neighbourhoods

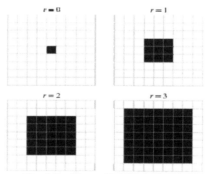

Fig. 2-8   Moore Neighbourhoods

A square-shaped neighbourhood that can be used to define a set of cells surrounding a given cell (x_0, y_0) that may affect the evolution of a two-dimensional cellular automaton on a square grid the Moore neighbourhood of range r is defined by

$$N^M_{(x_0,y_0)} = \{(x, y): |x-x_0| <= r, |y-y_0| <= r\}$$

Moore neighbourhoods for ranges r=0, 1, 2, and 3 are illustrated above. The number of cells in the Moore neighbourhood of range r is the odd squares $(2r + 1)^2$, the first few of which are 1, 9, 25, 49, 81, ... etc. In this case, for d-dimensional square meshes with a neighbourhood range r=1, the number of neighbours for the cell can be calculated to be (3^d-1).

(3) Extended Moore Neighbourhoods

By extending the neighbour radius r to larger than 2 or even more, one obtains so-called extended Moore type neighbours. The number of neighbours in d-dimension on square meshes with Extended Moore type can be expressed as $((2r+1)^d - 1)$.

(4) Margolus Neighbourhood

In the above neighbourhood schemes, every cell is treated as standalone element, evolving synchronously with other integrated elements. Margolus neighbourhoods are different, since transitions are applied to cells found in non-overlapping 2-cell blocks. (Such as 2 × 2 squares in two dimensions, or 2 × 2 × 2 cubes in three dimensions, etc.) The blocks are shifted by one cell (along each dimension) on alternate time steps. This simple partitioning scheme turned out to be very useful for modelling physical systems. Another important property of a Margolus neighbourhood is that it allows for a very easy creation of reversible rules at the microscopic scale, as elaborated on (http://cell-auto.com/neighbourhood/margolus/).

     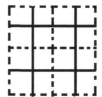

Fig. 2-9 (a) Margolus neighbourhood at t   **(b)** Margolus neighbourhood at (t +1)

2.4 Transition rules of CA

Cellular Automata update the state values of all cells synchronously at discrete time steps according to a transition rule R. Such rule is based on the state of the cell itself and on the state of its neighbours. There are several kinds of evolution rules (Wolfram, 2002; Ganguly et al., 2003).

2.4.1   Deterministic CA rules

The rules of cellular automata in general are deterministic in nature (Burks, 1970; Wolfram, 1984; Karafyllidis & Thanailakis, 1997; Chen et al, 2002) similar to spatial discretization of differential equations. They are said to be deterministic if a given initial value of the state variable yields a unique value of the state variables at the next time level. Take e.g. the "parity rule" for the 5-neighbourhood:

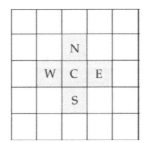

Fig. 2-10 "parity rule" for the 5-neighbourhood

If the state variable is Boolean or Logical type, the CA rule could be described as:

C_i^{t+1} = mod {($(C_i^t + N_i^t + E_i^t + S_i^t + W_i^t$), 2}, viz.
C_i^{t+1} = 1 if the sum of the neighbourhood states (including C itself) is odd, and
C_i^{t+1} = 0 otherwise.

For other types of state variables, the Parity Rule is simply the average value among the total 5-neighbourhood

C_i^{t+1} = ($C_i^t + N_i^t + E_i^t + S_i^t + W_i^t$) / 5

In the "Parity Rule" all cells are treated the same way, which is why it is very popular in 2-dimentional cellular automata and most often used in deterministic CA rules.

2.4.2   Probabilistic CA rules

There are variations, however, in which the rule sets are probabilistic, or fuzzy. They are said to be probabilistic if the value of the state variables at the next time step is conditioned by a probability distribution or random variable, which is compared to a pre-determined numerical value that is a function of the initial state variable (Guinot, 2002).

One example is the "Voter Rule" for the 5-neighbourhood scheme (here we take a Boolean state variable as an example):

C_i^{t+1} = 1   if ($N_i^t + E_i^t + S_i^t + W_i^t$) > 2
C_i^{t+1} = 0   if ($N_i^t + E_i^t + S_i^t + W_i^t$) < 2
C_i^{t+1}= ~C_i^t otherwise

where ~C is the complement of C:
~C= 1   if C = 0;
~C= 0   if C =1

It was observed that in the "Voter Rule" the results may more depend on the value of C_i^t and it's locally neighbours, as compared with the "Parity Rule".

2.4.3   Data-driven based rules

There are more cellular automata models that apply stochastic rules in which the probability of transition is calculated and generated a random process, simulating evolution. (Chen & Mynett, 2003a; Bandman, 2002). Data-driven approaches can be used to generate the rules for cellular automata. There are several kinds of data-driven based CA rules, such as ANN based rules and Fuzzy Logic rules. Fuzzy Logic rules in cellular automata have the advantage that (i) fuzzy sets and cell states are both finite (Flocchini, et al., 2000); (ii) inference in fuzzy logic and evolution in CA are rule based; and (iii) empirical knowledge is easily incorporated (Chen & Mynett, 2004).

2.4.4   Asynchronous rules

In most cases, the state of a cell depends upon the output from the precious state. However, there are some time-independent rules. For example, two alternate rules at even and odd time steps were used in the problem of directed percolation (Chopard & Droz, 1998).

2.5 Boundaries of CA

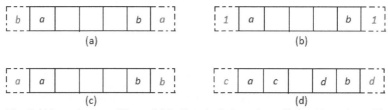

(a)   (b)

(c)   (d)

Fig. 2-11 boundary conditions of CA: (i) periodic boundary; (ii) fixed boundary; (iii) adiabatic boundary; (iv) reflection boundary (dashed lines represent the virtual cells).

In theory, cellular space in each dimension extends to infinity. However, in practical implementations, this ideal condition cannot be achieved, so we need to define different boundary conditions to deal with the edges of the domain. There are four main types of boundary conditions (Fig. 2-11): (i) periodic; (ii) reflective; (iii) adiabatic and (iv) fixed. Sometimes, in real applications of more realistically simulating natural phenomena, there may exist random type boundaries, i.e. a random value is generated in real time at the boundary. (Chen, 2004)

(i) Periodic Boundary. This refers to a virtual boundary 'connecting' the end-cells. For the one-dimensional case, this leads to a 'circle'. For a two-dimensional CA space, periodic boundaries connect the upper and lower, as well as left and right edge (Fig. 2-12). The shape of topological torus looks like a 'tire' (Fig. 2-13). Periodic type boundary conditions are closest to infinite space and are often used in CA-research.

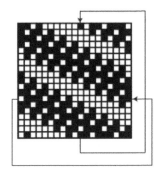   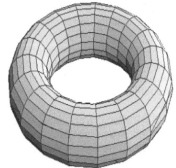

Fig. 2-12 Neighborhood Wrap-Around   **Fig. 2-13** Topological Torus

(ii) Fixed (Constant) Boundary. For the entire domain the outer boundary value is taken to be a fixed pre-assigned value, such as 0, 1, and so on.

(iii) Adiabatic Boundary. The virtual cells beyond the border are assumed to have the same value as the boundary cell.

(iv) Reflection Boundary. The state of the cells outside the border is determined by reflection (setting the boundary as the 'mirror-axis') from the neighbour inside the border.

Typical for CA based ecosystem modelling, the fixed boundary conditions are often used for aquatic ecosystem, while the adiabatic boundary condition is usually used for the terrestrial ecosystem modelling (Chen, 2008). It should be noted that these four types of boundaries in practical applications, especially in two-dimensional or higher dimensional configurations, can be combined with each other. Often in two-dimensional space, the upper and lower boundaries use the reflection type; while the left and right boundaries use periodic type.

2.6 Behaviour and classification of CA

Classification of cellular automata is an important research topic at the core of cellular automata theory. Based on different starting points, there are a variety of classification categories of cellular automata, among them the most influential classification was undoubtedly done by Wolfram in the early 1980s. Wolfram (1984) classified cellular automata based on their dynamic behaviour. Gutowitz proposed a hierarchical and quantitative classification system based on the behaviour of cellular automata with Markov probability (Gutowitz, 1991). In this section, we give a further description and survey about the two classification categories.

Wolfram (1984) began a systematic investigation into the evolution behaviour from one-dimensional cellular automata during the early 1980s. The one-dimensional CA has two neighbourhoods (left and right) and two states, which is possibly the simplest CA configuration. The lattice in these 1-d CAs is a line, and cells are updated based on their own state and their immediate neighbours. As the neighbourhood size is equal to 3, and the number of states is equal to 2, in total 256 (2^{2*2*2}) CA rules are possible. Wolfram identified these 256 kinds of rules (from 0 to 255) for Elementary Cellular Automata. Based on a large number of computer experiments, the kinematic behaviour of all cellular automata was grouped into four categories (Wolfram, 1986):

(1) Class I--- Homogeneous type: CAs evolve to a uniform configuration from any initial state, which means after a certain runtime CAs stabilize in cellular space, where the space remains steady in a fixed state and does not change with time. This state can be thought of in dynamical systems terms as a 'point attractor', or 'limit point'.

(2) Class II---Periodic type: After running a certain period of time, a series of cells tends to a fixed structure (Stable Patterns) or periodic structures (Periodical Patterns). Since these structures may be seen as a filter, the approach can be applied to image processing. The evolution of Class II CA with periodic configurations can be thought of as analogous to 'limit cycles' in dynamical systems.

(3) Class III---Chaotic type: from any initial state, after sufficient runtime, cellular automata exhibit a non-periodic or chaotic behaviour, comparable to fractal dimension features. Small changes in initial lattice configuration can lead to larger and larger changes in resulting configurations, as is the case is for chaotic dynamical systems.

(4) Class IV---Complex type: in some sense Class IV is 'between' the purely chaotic behaviour of Class III, and the static behaviour of Class II. These types of CA exhibit propagating structures, where the emergence of complex global structures from initial local disturbances, will continue to spread.

From the perspective of research on cellular automata, most valuable research is the fourth class, because such class can be considered to have the virtue of "emergence computation", which can be used as a generalized computer (Universal Computer) to simulate arbitrarily complex calculation processes. In addition, this type of cellular automata in the development process also shows a strong irreversible characteristic.

Wolfram also described the approximate probability of four classes' cellular automata. It was noted that the CA with complex patterns appears with relatively smaller probability, while the third type of CA with chaotic configurations occurs more frequently, and the probability shows an increasing trend when k (the finite state number) and r (neighbour radius) increase.

Although the above classification is not a strict mathematical classification, Wolfram was able to identify many types of dynamical behaviour from only these four classes of cellular automata, which is a very meaningful finding and has great significance for cellular automata. This classification may be universal, so that it is likely there are many physical systems or living systems can be classified according to this method. Although the details may be different, their behaviour is similar.

Another classification method is based on the number of dimensions of cellular automata. Theoretically, cellular automata can have any number of dimensions. So, according to the dimension of cellular space classification, cellular automata can be divided into (Wolfram, 2002).

(1) One-dimensional cellular automata.
One-dimensional cellular automata were researched systematically at an early stage. Generally speaking, it has relatively simple rules, and all possible rules can be studied in-depth. Wolfram classified the dynamics of cellular automata based on the analysis of elementary one-dimensional cellular automata (Elementary Cellular Automata). The distinguished features of Elementary Cellular Automata are easily visualized to observe their dynamic evolution: as spatial-temporal visualization.

(2) Two-dimensional cellular automata.
Here the two-dimensional Euclidean space is divided into lattices, like in Conway's most widely used "Game of Life" (Gardner, 1970). Because many phenomena in the world show a two-dimensional distribution, and there are some other phenomena that can be converted to two-dimensional space by abstraction or mapping methods, the application of two-dimensional cellular automata is most widely used and very popular. Many applications in CA modelling are based on two-dimensional cellular automata.

(3) Three-dimensional cellular automata.
(Bays, 1988) carried out a number of experiments in this respect, including achieving the Game of Life in three-dimensional space, which is the continuation and expansion of one-dimensional and two-dimensional cellular automata theory.

(4) Higher-dimensional cellular automata.
Currently, only a few theoretically investigations exist but lack actual system modelling in real applications. Meeker (1998) explored four-dimensional cellular automata.

2.7 Applications of Cellular Automata

Since its introduction, CA has been widely applied to various fields of computer science, mathematics, physics, chemistry, biology, ecology, sociology, geography, environmental science, information theory, and military science.

In computer science, Cellular Automata can be seen as a parallel computer and are used to study parallel computing (Wolfram, 1983). In addition, Cellular automata are also used in computer graphics research.

In mathematics, Cellular Automata are used to study number theory and parallel computing. For example, (Fischer, 1965) designed prime number filters (Prime Number Sieves). (Wolfram, 1983)

In physics, the lattice gas cellular automata were most successfully applied in fluid mechanics. CA simulation is also applied to magnetic field theory, electric fields, etc., as well as to thermal diffusion, thermal conductivity and mechanical analog waves. In addition, CA are also used to simulate snowfall, avalanches, and dendrite formation.

In chemistry, cellular automata are used to study chemical reactions and interactions by simulating atoms, molecules, and other microscopic particles in a chemical reaction. (Bar-Yam, 1996; Ostrovsky et al., 2001) used cellular automata to simulate the polymerization process.

In biology, since cellular automata are derived from the idea of biological self-reproduction, its application is more natural and widespread. Cellular Automata are used on growth of tumor cells, exploring the mechanism of the human brain (Victor, 1990), HIV infection process (Sieburg, 1990), and self-reproductive biological phenomena such as the latest popular clone technology (Ermentrout, 1993).

In ecology, CA were used for population dynamic process simulation, such as rabbit--grass, shark-fish interactions etc., which demonstrated satisfactorily the dynamic effects; cellular automata were also successfully applied to ants, geese, migratory fish and other animals simulating group behaviour. In addition, the CA-based biomes dispersion model has currently become to be a hot item.

In sociology, cellular automata are used to simulate formation and outbreak processes of economic crisis, social behaviour of individuals, popular phenomena (such as clothing fashion, colour preference, etc.).

In Geography, Tobler (1970) proposed the use of cell space models for modelling geographic interactions, but CA matured as a unban technique tool around the end of the 1980s. Green et al. (1989) used CA as a generic tool for modelling land-scope dynamics. Itami (1994) showed cell-based GISs may indeed serve as a tool for implementing cellular automata models for the purposes of geographic analysis.

In information science, Cellular Automata are used for information storage, image processing and pattern recognition (Deutsch, 1972), (Sternberg, 1980), (Rosenfeld &Wu, 1980).

In environmental science, researchers use cellular automata to simulate oil spill behaviour (Ha et al., 2012), waste water pollution, gas dispersion, and other diffusion processes (Slimi & Yacoubi, 2009).

In military science, cellular automata are used for military combat simulation and battlefield understanding (Lauren, M. K., 2001).

Chapter 3

The concept of Unstructured Cellular Automata

3.1 Motivations to develop Unstructured Cellular Automata

Cellular Automata (CA) were originally associated with regular grids, or more particularly, rectangular cells on a structured grid. Classical CA assumes that the structure of the cell configuration and the number of neighbours are homogenous for every location in cellular space. Since geographical features in nature (like rivers, coasts or lakes) are usually not regularly shaped, classical Cellular Automata with identical shape and size are not the most adequate to represent complex geometries in real world. In order to overcome this limitation, some researchers have explored to extend Classical Cellular Automata to irregular cell configurations (Flache & Hegselmann, 2001).

In this thesis, the concept of Unstructured Cellular Automata (UCA) is developed for unstructured computational meshes which allow more modelling flexibility missing in structured grids. Allowing variable sizes of elements, permits a more accurate representation of boundaries without requiring an excessive number of points and elements. In unstructured cellular automata, the state of a cell depends upon the state of the cell at the previous time step as well as the states of the cell's immediate neighbours. The corresponding evolution process could be described as

$$a_i^{t+1} = \phi\left(a_i^t, a_j^t, a_{k,}^t \ldots\right)$$

where φ is a function of local evolution rules, the subscripts i, j, k, ... refer to the positions of the neighbouring cells, and superscripts t, t+1, ... refer to the cell states at different time intervals. All cell states are updated synchronously in discrete time steps according to local evolution rules.

3.2 Cell configurations of Unstructured Cellular Automata

3.2.1   Unstructured Cellular Automata with triangle elements

Fig. 3-1(a) shows the basic scheme of structured cellular automata, together with the triangle-based unstructured cellular automata as in Fig. 3-1(b); three neighbours of the triangles were taking into account. The neighbours share the common edges with the central triangle. The state of the central triangle is determined by the previous state and the state of its three neighbours.

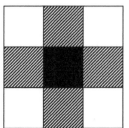
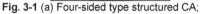

Fig. 3-1 (a) Four-sided type structured CA;   (b) Three-sided type unstructured CA

Wolfram (1983) pointed out that for a one-dimensional classical cellular automaton with two neighbours and two states, there are $2^{2*2*2} = 256$ possible transition rules. In case of two-dimensional unstructured cellular automata, taking the simplest configuration consisting of triangular elements as an example, each element now has three neighbours as shown in Fig. 3-1(b). Suppose each cell has three possible states, Red-Green-White, say, then the total number of possible rules could be calculated as $3^{3*3*3*3} = 4.4*10^{38}$ which is a huge number so it is impossible to explore all rules.

In this chapter, we use the fair rule, which looks like 'Rock-Paper-Scissors', which also called cyclic cellular automaton (Fisch, 1992). In this system, each cell remains unchanged until some neighboring cell has a special value exactly one unit larger than that of the cell itself, at which point it copies its neighbor's value.

$$a_i^{t+1} = \begin{cases} 1, & if \ a_i^t = 0, \ a_{ij}^t = 1 \\ 0, & if \ a_i^t = -1, \ a_{ij}^t = 0 \\ -1, & if \ a_i^t = 1, \ a_{ij}^t = -1 \\ a_i^t, & \text{otherwise} \end{cases}$$

where,

a_{ij}^t is anyone of the neighbours of a_i^t

$j = 1,2,3$ for Three-sided type unstructured CA.

Representing $a_{ij}^t = -1, \ 0, \ 1$ with three colours (green, white, red). In this thesis, three colours rule of UCA implies: "Green" beat "White"; "White" beat "Red"; "Red" beat "Green".

Running this rule in simulation for 100 time steps leads to the distribution as shown in Fig. 3-2.

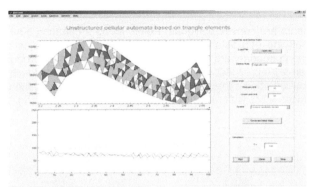

Fig. 3-2 Unstructured CA based on triangular elements implemented with 'Three-sided' rule.
(Initial distribution Red33% Green33% White33%)

It should be noted there are two sub-figures in this Figure: the upper one displays the spatial distribution pattern (at the last time step), while the lower lines with different colours represent the population dynamics in time. The results from Fig. 3-2 show a dynamic stability. No obvious patchiness patterns can be observed and the populations fluctuate around some seemingly stable values.

3.2.2   *Unstructured Cellular automata with polygon elements*

In the previous section, unstructured cellular automata were introduced taking one triangular element at the centre, in analogy with the classical Von Neumann arrangement in structured CA. During trial experiments it was observed that the difference between taking a common-vertex neighbour and a common-edge neighbour as neighbouring cells affect the state of the central element. In this section, we try to explore a scheme that avoids such behaviour, by taking the common vertex between cells as the centre. In Fig. 3-3, the Voronoi based Cellular Automata is shown as polygons around cell vertices. The triangular elements in the Voronoi diagram (blue lines) are the triangular elements of the grid, while the resulting Voronoi polygons are indicated in red (Flache & Hegselmann, 2001).

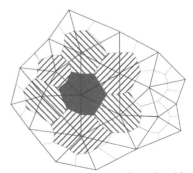

Fig. 3-3 Unstructured cellular automata based on Voronoi diagram
(the red lines indicate the Voronoi polygons)

Voronoi elements are convex polygons. Points along a common boundary between Voronoi regions are equidistant to the corresponding spatial objects. Objects that share a common boundary are neighbours to each other in the Voronoi spatial model (Aurenhammer & Klein, 2000; Carvalho et al., 2002). Compared with the triangular scheme introduced above, the polygon scheme effectively places the state variables in the cell-vertices (CV), with related vertices as 'neighbours' (see Fig. 3-4).

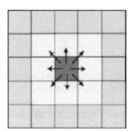 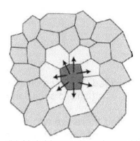

Fig. 3-4 (a) Neighbours in structured CA.   (b) Neighbours in polygon UCA

From Fig. 3-4 the analogy can be seen with a Moore scheme on a rectangular grid. More in general, it has been found that the number of neighbouring cells can vary from 3 to 14 in a Voronoi graph, while in case of structured CA the number of neighbouring cells is constant (Flache & Hegselmann, 2001).

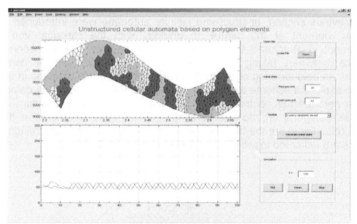

Fig. 3-5 Unstructured CA simulation based on Voronoi polygon elements
(initial conditions Red 33%, Green 33%, White 33%)

From Fig. 3-5 it is observed that the simulation result shows that the populations become stable quite quickly, and patchiness appears. In the cases explored here, the polygon-based unstructured cellular automata always showed advantages with respect to stability.

3.3   Effects of Initial Conditions in UCA

CA evolutions are in principle irreversible, so initial conditions of cellular automata are usually set at random. However, in modelling practice it is observed that initial conditions can influence the macro-level spatial structures that emerge from local interactions. So the effects of initial conditions in UCA are worthy of further elaboration. In reality, observations from the real world could be obtained from a geographical information system (GIS) or from remote sensing (RS) data (Chen, 2009). Such set of high solution images can be processed to provide as the initial conditions for (U)CA modelling, as demonstrated in an application later in thesis (Chapter 7).

3.3.1   Initial spatial distribution

3.3.1.1   Three colours randomly mixed

If a uniform distribution is assumed, an initial spatial distribution can be randomly generated. This is referred to hereafter as "three colours random mix". It is observed that after some initial transitions, the populations exhibit only mild fluctuations around a constant value (the evolution lines in the Figures below).

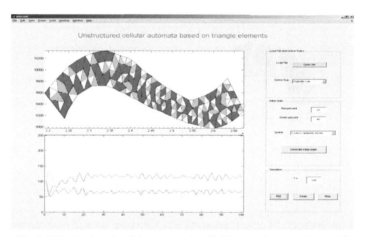

Fig. 3-6 Unstructured cellular automata with "three colours random mix"
(Red 20%, Green 40%, White 40%)

21

3.3.1.2 One colour isolated near a boundary

Fig. 3-7 (a) shows the initial condition when all 20% red elements are located on one side, while the 40% green elements and 40% white elements are randomly distributed over the remaining space.

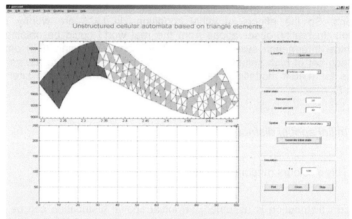

Fig. 3-7 (a) Initial conditions with "one colour isolated near one boundary"
(Red 20%, Green 40%, White 40%)

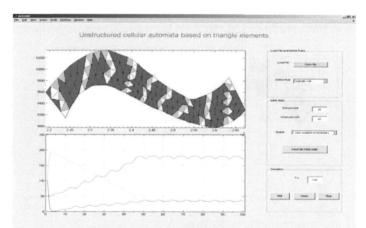

Fig. 3-7 (b) Unstructured cellular automata with "One color isolated near one boundary".
(Red20% Green40% White40%)

Figure 3-7 (b) illustrates the result when 20% red elements isolated near one boundary. From the simulation, the following phenomena are observed. At the beginning, since the red ones were isolated, the green ones ate most white ones, causing the green population increasing suddenly. Meanwhile, the white ones decreased quickly. Later, with red ones propagating, the green population reduced and the red number rise until these three colors reached an equilibrium state.

3.3.1.3   One colour isolated in the middle

Fig. 3-8 (a) shows the initial condition when all 20% red elements are located in the middle; Fig. 3-8 (b) shows the result that the system reaches a stable state after 30 steps when the red elements are isolated in the middle.

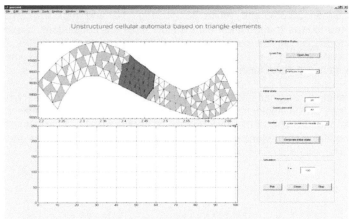

Fig. 3-8 (a) Initial distribution with 20% red isolated in the middle
(Red 20%, Green 40%, White 40%)

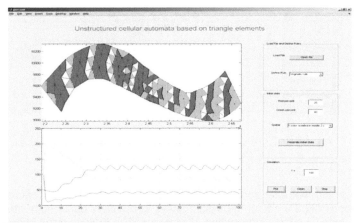

Fig. 3-8 (b) Unstructured cellular automata with "red isolated in the middle".
(Red20% Green40% White40%)

Comparing Figure 3-6 with Figure 3-7(b) and Figure 3-8(b), it is very clear that the propagating speed depends on the initial distribution. For the randomly mixed case, the red number increases very quickly; and the system reaches the stable state after 20 steps. For the middle isolated case, it takes 30 steps to stabilize. The boundary-isolated case is slowest, which requires 60 steps before the stable state.

3.3.2   Initial percentage analysis

Initial conditions are two-fold: initial spatial pattern and initial percentage distribution. In the previous section, the initial spatial pattern was considered. In this section, the initial percentage is analysed. In the previous sections already two different initial-percentages were considered. The first case was the equal distribution Red 33.3%, Green 33.3%, and White 33.3% (Fig. 3-2 and Fig. 3-5), which can be considered to lead to the fairest competition since the populations of species at the beginning are equal. The second case with the percentages "Red 20%, Green 40% and White 40%" (Fig. 3-6, Fig. 3-7 and Fig. 3-8 resp.) was an situation where only two species have the same initial population.

In this section, another case is analysed, where all three species have different populations at the beginning. We take "Red 20%, Green 30%, White 50%" as an example. Fig. 3-9 shows the performance of unstructured cellular automata with the initial conditions characterized by "three colours random mix" in space and "Red 20%, Green 30%, White 50%" in population percentage.

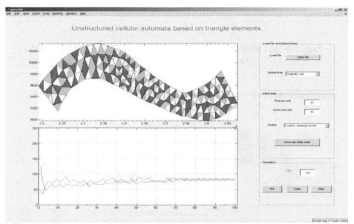

Fig. 3-9 Unstructured cellular automata with "three colors random mix"
(Red 20%, Green 30%, White 50%)

It is observed that after some 70 running steps, a stable state is reached of equal distributions (33.3% each); the system stays in a steady equilibrium after that. The initially unequal population situation disappears. The three species keep competing with each other leading to a dynamic equilibrium. However, the results turn out be totally different if other initial distributions are considered, i.e. "one (red) is isolated on a boundary" and "one (red) is isolated in the middle", as depicted in the following two figures below.

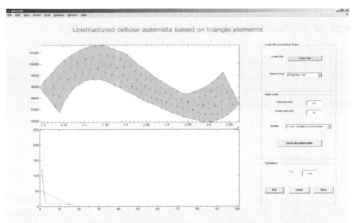

Fig. 3-10 UCA with initial condition "one type isolated on boundary"
(Red 20%, Green 30%, White 50%)

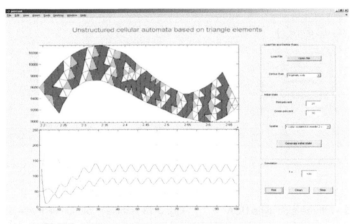

Fig. 3-11 UCA with initial condition "one isolated in middle'
(Red 20%, Green 30%, White 50%)

From Fig. 3-10 it can be observed that with 20% red clustering at the boundary, the green becomes the dominant survivor and the other two disappear. From Fig. 3-11 it can be seen that the outcome becomes somewhat dynamically stable again after multiple time steps. From this and several other experiments, the conclusion can be drawn that only when the initial distribution is reasonably uniform, the results become stable after initial transition; otherwise, one of species may become the winner, while the other two will die out.

More UCA simulations with triangle elements were demonstrated in **Appendix-A**.

3.4 Effects of different Neighbourhood Schemes in UCA

The concept of unstructured cellular automata developed in this thesis may be more flexible compared with classic cellular automata; it also leads to much more complex problems of neighbourhood schemes than classic CA. In this section some considerations on neighbouring schemes of UCA are put forward. (Lin & Mynett, 2010)

3.4.1   Three-sided type

Fig 3-12 shows the basic scheme of unstructured cellular automata based on unstructured meshes as compared with structured cellular automata; three neighbouring triangles were taking into account. The neighbours share the common edges with the central triangle. The state of the central triangle is determined by its previous state and the state of its three neighbours, i.e.

$$a_i^{t+1} = \phi\left(a_i^t, a_{i1}^t, a_{i2}^t, a_{i3}^t\right) \qquad\qquad (3\text{-}1)$$

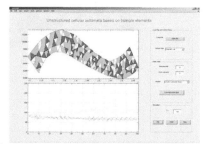

Fig. 3-12 'three-side' type UCA   **Fig. 3-13** sample evolution of 'three-sided type'

Using simple evolution rules (section 3.2.1) and a random initial distribution of 'green', 'red' and 'white', a distribution pattern evolves after a simulation of 100 time steps as shown in Fig.3-13. The upper figure shows the spatial distribution pattern where no obvious patchiness can be observed. The lower curves with corresponding colours represent the time evolution of the individual species (often referred to as population dynamics), which show dynamic equilibrium after multiple time steps. The stabilized populations of different species also seem random.

3.4.2   Moore type UCA

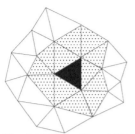

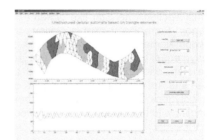

Fig. 3-14 'Moore' type UCA   **Fig. 3-15** sample evolution of 'Moore' type UCA

The Moore type considers more neighbours of the central element. For a structured grid, a Moor type configuration has 8 neighbours (Fig. 3-4a), while for unstructured grids, the number of neighbours is undetermined (see Fig. 3-14). However, if we consider common vertices with the central element, we obtain a neighbourhood configuration similar to Moore's for structured grids. The results shown in Fig. 3-15 indicate that the population reaches a stable state in this case. It should be mentioned that a patchiness pattern can appear. This can be contributed to the fact that more neighbours are contributing when sharing common vertices with the central triangle.

3.4.3   Three-vertex type

Since every triangular element has three vertices, we can consider the contribution of three vertices (vertex 1, vertex 2, and vertex 3) to the central element (see Fig. 3-16). For every vertex, the state is determined by the triangles surrounding it. For instance, the vertex 3 is surrounded by triangle 1, 4, 5, 6 and 3.

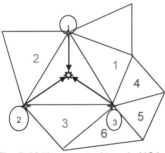

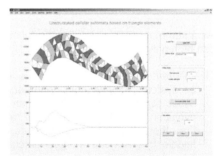

Fig. 3-16 'three-vertex' type in UCA   **Fig. 3-17** sample evolution of 'three-vertex' type

In fact, we can take the same numbers of neighbours into account as we do in the Moore type. But in this case, "common-edge" and "common-vertex" neighbours can be distinguished. In Fig. 3-16, triangles 1, 2 and 3 are "common-edge" neighbours, while triangle 4, 5 and 6 are "common-vertex" neighbours. The "common-edge" neighbours are more dominant than the "common-vertex" neighbours. The results in Fig. 3-17 show that in the beginning there are some fluctuations, but after a few time steps, the population reaches an equilibrium state.

3.4.4   Star-like Voronoi polygons

In this section, the Star-like scheme is proposed which is based on the Voronoi polygon. From Fig. 3-18 it is obvious that no matter how the shape and size of the polygon elements vary, the area of the central polygon element with the red colour is always equal to the total area of the green triangular neighbours. For every discrete time step, each polygon is executed as a cell and its common-edge triangles are taken into account as the neighbour's cells of the central polygon. Meanwhile, when the polygons are regard as neighbourhoods, they can be divided into several triangles with the same status who affect their neighbouring polygons.

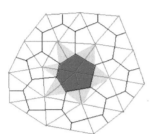

Fig. 3-18 'Star-like' type in UCA

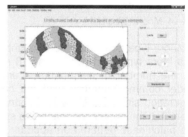

Fig. 3-19 Sample evolution of 'Star-like' type

From Fig. 3-19 it can be seen that the result shows the populations to become stable rather quickly, but also patchiness appears. Compared with other schemes, this type of unstructured cellular automata modelling has advantages mainly in stability.

More UCA experiments with polygon elements were demonstrated in **Appendix-B**

3.5 Analysis and discussion

As a summary of this chapter, the following conclusions from the simulation studies can be highlighted.

(1) *Different initial spatial distributions*: when three species are randomly mixed, the ecosystem reaches equilibrium quickly. Among the different randomly mixed cases considered here, either when all the three species are randomly mixed the evolution of the populations quite rapidly reaches a dynamically stable state. On the other hand, when one species is isolated either on the boundary or in the middle, this may cause bigger patchiness or even lead to the dominant solution of one species

(2) *Different initial percentages*: if there exist more fair distributions at the beginning (e.g. 33% for each of the three species), the UCA shows more active equilibrium characteristics. If there are two kinds of species with the same initial population, and if there is no big difference between this number and that of the third species, the ecosystem also can reach an equilibrium state; otherwise, either one or two species will die out, or the system becomes unpredictable.

(3) *Polygon element*: the polygon element has the advantage of balancing the equilibrium conditions and manifesting the occurrence of patchiness characteristics. It also takes more neighbours into consideration. Compared with triangle-based unstructured cellular automata, this kind of paradigm shows bigger patchiness than triangular cell-based configurations. In addition, whatever the initial conditions are, the results converge to a steady state, where the numbers of different species are almost same

(4) *Neighbouring schemes:* it was observed that the 'Three-sided type' is most stable, while the 'Moore' type and 'vertex type' show clearer patchiness. This is because the 'Three-sided type' is based on few local rules, which usually leads to quasi-stable dynamic states. In 'Moore' type and 'vertex type' configurations, more neighbours are taken into account and more patchy patterns appear. In order to balance the equilibrium and the patchiness characteristics, and to take more neighbours into consideration, the Unstructured Cellular Automata based on polygon-type elements were developed. Compared with triangle-based unstructured cellular automata, this kind of paradigm shows more patchiness but is more stable than 'Three-Vertex' type configurations. Whatever the initial condition is, the result is always stable, with the population of different species converging to their average value. When unstructured cellular automata are used in practical applications, the evolution rules of UCA should be selected based on real-life features of the phenomenon considered.

Chapter 4

Computational Theory of Unstructured Cellular Automata

4.1 Cellular Automata relations with other modelling methodologies

The development of Cellular Automata and related methodologies are inextricably linked. On the one hand, CA builds on theories such as logical mathematics, discrete mathematics, Turing machine; on the other hand, the development of CA also contributed to a number of related disciplines and modelling paradigms (such as Artificial Intelligence, nonlinear systems, complexity theory), and even directly led to the generation of artificial life sciences. (Wolfram, 2002)

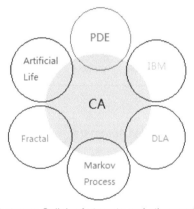

Fig. 4-1 Links between Cellular Automata and other modelling paradigms

Fig. 4-1 shows several different methodologies that are widely used nowadays and their relation to CA. Some have a long history, e.g. Partial Differential Equations (PDEs) and Markov processes (Markov, 1907), but others have recently emerged with the advent of digital computers, notably Artificial Life (Langton, 1986); Fractal Theory (Mandelbrot, 1967); Diffusion-Limited Aggregation (DLA) (Written & Sander, 1983); and Multi-Agent system (Ferber, 1999)

In this section, five modern computational methods (Artificial Life, Fractal Theory, Markov Chains, Diffusion-Limited Aggregation and Multi-Agent Systems, see Table 4-1) will be compared with cellular automata methods. Their relationships including similarities and differences are briefly discussed. Section 4.2 will pay special attention to the relationship between CA and PDEs. In Chapter 7, the study case of UCA will be described and compared with Individual Based Modelling (IBM).

Table 4-1 Popular methodologies compared with CA

Modelling paradigms	*Similarity* with CA	**Difference** with CA	
		CA	modelling paradigm
Artificial Life (AL)	Local interactions create global behaviour in non-linear dynamic systems	CA is an important branch and research tool of AL	AL was based on in-depth research of CA
Fractal Theory	Both make use of repetitive local interactions to create global patterns	CA focuses on a specific mechanism of interaction, from which global patterns emerge	FT has statistical implications when deducing the overall characteristics from local structure
Markov Processes	dynamical model with discrete time and discrete state	CA states are closely related to MP concept of spatial location	there is no concept of space in Markov Chains only state variables
Diffusion Limited Aggregation modelling (DLA)	Both are discrete in space and time and can generate similar complex patterns	CA covers the entire grid space; each cell can only have one active state at a time, states are changing;	DLA considers the movement of specific particles that can have variable states; multiple particles can occupy one grid cell.
Individual Based Modelling (IBM)	Both are discrete in space, focus on the interaction between individuals creating complex global behaviour	IBM usually comprises individuals with sparse distribution, and only calculates individual behaviour	CA covers the entire grid and calculates the status of each grid cel among the entire domain

4.1.1 Artificial Life and Cellular Automata

Artificial Life (AL) emerged in 1987 and, although still relatively new, is one of the important complexity science disciplines. AL can mimic some characteristic behaviour of natural living systems such as self-replication, parasitism, competition, evolution, collaboration. Besides, Artificial Life can simulate "a possible phenomenon of life" (Life-as-it-could-be), so that people can deepen their understanding of the "known phenomenon of life" (Life-as-we-know-it) (Langton, 1995).

A cellular automaton is an important branch and research tool of artificial life. Langton (1986) proposed the concept of artificial life based on in-depth research on cellular automata. At the same time, the development of artificial life gave a new meaning to cellular automata. CA modelling received renewed understanding and recognition, and in the 1990s the theory of CA and its applications were further improved.

CA has considerable similarities with AL modelling. Cellular Automata and other artificial life models (such as neural networks, genetic algorithms and the like), are both based on simplified local interactions that, together and at a larger scale, turn into an overall complex system behaviour. Furthermore, cellular automata, neural networks, and artificial life systems can be classified as network dynamics models that belong to non-linear dynamics; they are closely related and interrelated with each other. Recently a product called cellular neural network (CNN) was introduced as a combination between cellular automata and artificial neural networks.

4.1.2   Fractal Theory and Cellular Automata

Cellular automata and fractal theory are closely linked. The repetitive application of self-replicating algorithms in case of cellular automata often leads to self-similar fractal configurations in space, which means that the performance of cellular automata sometimes can even be quantitatively described by fractal theory. The classic example of fractal theory in itself resembles a cellular automaton model. However, there are also essential differences between cellular automata and fractal theory: Cellular Automata focus on representing the underlying mechanism of the phenomenon, while Fractal Theory studies the expression of the phenomenon.

CA usually starts from the rule of the particular phenomenon, and then creates a model to simulate the evolution of the phenomenon. A Fractal Model on the other hand, need not be based on physical or laws, but can be based on a mathematical algorithm that is applied repetitively to create particular complex phenomena by multiple self-similarity as described by its fractal dimension. Cellular automata and fractal theory both evolve from the local to the global level, but in essence there is a huge difference between them. The self-similarity of Fractal Theory has statistical meaning when deducing the overall characteristics from the local structure. Contrary to that, the essence of cellular automata is that "emerging" properties of complex behaviour of the structure as a whole, although derived from certain simple local rules, leads to features at the macro-level that local-levels do not have.

In essence, fractal theory emphasizes the similarity and relevance between local and global behaviour, while cellular automata focus on "emerging" overall behaviour, including uncertainty and the nonlinear relationships between the global behaviour and the local structure.

4.1.3   Markov Processes and Cellular Automata

Markov Processes are stochastic processes where the state variables time and space may either be continuous, or discrete. Markov processes with discrete time steps and discrete states are called Markov chains (James, 1998). Markov chains and cellular automata are both discrete dynamic models, and have a number of conceptual similarities. Especially for stochastic cellular automata, each cell can be viewed as a Markov chain with no backward effect in time and no external effect in space.

But even for stochastic cellular automata there are considerable differences when compared with Markov chains. First of all, there is no concept of space in Markov chains, only state variables that evolve in time, while the states of cellular automata are closely related to their spatial location. Secondly, the transition probability of states in a Markov chain is often pre-configured, while the transition probabilities of stochastic cellular automata are decided by the state of the current configuration and its neighbours' states.

4.1.4   Diffusion-Limited Aggregation model (DLA) and Cellular Automata

A Random Walk Model (from Wikipedia: Pearson, 1905) is a statistical mathematical model which is commonly used to provide "the most likely state". Random walk models usually include a number of particles that follow the same rules but with different random parameters, so their movements become independent of each other. If interactions are considered, it is possible to construct other models based on random walk, such as the cohesion–diffusion model.

Written and Sander (1983) first proposed the Diffusion–Limited Aggregation (DLA). DLA can be seen as a multi-particle random walk model and its computational domain often includes a discrete grid. Voss (1984) improved DLA and explored the Multi-Particle Diffusion Aggregation model.

Like cellular automata, random walk models and cohesion–diffusion models can generate similar complex patterns. But there are still some differences between them. Differences between random walk models and cellular automata are:
- the random walk model usually only considers the motion of individual particles, while cellular automata models include multiple cels;
- random walk models usually do not consider the interactions between particles, while CA models do;
- particles in random walk models follow the concept of movement, while elements in Cellular Automata models usually refer to a state change process;
- the space of particle movement in random walk models can be discrete or continuous, but cellular automata are on a discrete grid in space.

Multi-particle diffusion cohesion models are very similar to cellular automata: both of them are discrete in space and time; deal with interacting particles; effects have local features. However, there are still several points of difference between them: (i) a cellular automaton model contains the entire grid space, while cohesion diffusion models consider the movement of specific particles; (ii) cellular automata usually only have state changes of elements and their spatial positions are fixed, while the particle diffusion model not only changes in state, but also contains moving particles; (iii) cohesion diffusion models contain multiple particles which can simultaneously occupy one grid cell in space, while in cellular automata each cell can only have one active element. Thus, in a sense, the cohesion-diffusion model is more similar to the multi-agent model mentioned below; it can be seen as a "no mind" multi-agent model in which the particle has no particular objective and there is no competition and collaboration between particles.

4.1.5   Individual Based Model (IBM) and Cellular Automata

Multi-Agent Systems (MAS) is a hot topic in distributed artificial intelligence research (Shi, 2000) on collaboration, competition and other interactions between autonomous intelligent agents. Agent Based Models (ABM), also known as Entity Based Models (EBM) or called Individual Based Model (IBM) is a subset of multi-agent systems. Each agent in IBM represents a intelligent entity or individual in a real-world, like people in a crowd, individual plants and animals in ecosystems, a car in traffic flow, and so on.

Sometimes the agent in Individual Based Models is space-based, for example a vehicle in traffic flow, individual plants and animals in an ecosystem, etc. But some agents do not have any concept of space, such as in network computing. For space-based agents, space can be continuous or discrete. The cellular automaton model with discrete space is very similar to the space-based IBM: both are discrete in space, allow interactions between individuals, and result in a complex global behaviour. But still there is a big difference:

(1) Agents in IBM can move, e.g. individual animals, while elements in cellular automata are fixed in space;

(2) When IBM is based on a grid, the grid is just used to locate the spatial position of the agent, and several agents can occupy one grid cell; in cellular automata, each grid cell only represents one particular state;

(3) In essence, IBM usually comprises individuals with sparse distribution in space, while cellular automata cover the entire grid; IBM only considers individuals' behaviour, but cellular automata calculate the status of each grid cell among its entire space.

4.2 Cellular Automata and Partial Differential Equations

4.2.1   Analogies and differences between CA and PDEs

Partial Differential Equations (PDEs) have a history of several centuries. Famous scientists like Euler, Lagrange, Laplace, Poisson and so on, made many outstanding contributions. PDEs are the language of modern science used in many important fields of applied science. Typical features of PDEs are that they represent a space-time continuum. Contrary to that, cellular automata are fully discrete in space-time and in that sense; PDEs and CA can be seen as opposing calculation methods (Toffoli & Margolus, 1987).

The advantage of partial differential equations is that they can have accurate quantitative solutions, often in closed analytical form in case of simplified geometries. But modern digital computing is based on discrete representations of space-time co-ordinates leading to discrete solutions of PDEs, in close resemblance to CA. Still, CA models are different since (i) in CA the state variables are also discrete, providing the possibility to represent individuals rather than a continuum, and are therefore very well suited for population dynamics simulation; (ii) CA has very simple transition rules but can lead to rather complex behaviour patterns in both time and space (Guinot, 2002).

Furthermore, cellular automata modelling has the advantage of being easily understood, and being highly suitable for parallel computing, allowing the perspective of the model to multiple local views. However, continuum-based partial differential equations can easily be used to describe conservation principles in science and engineering and even in biological and ecological modelling (Bartlett and Hiorns, 1973). Recently, research has been done on using cellular automata as an alternative to (partial) differential equations. But when using a CA approach one has to face the problem of deducing the transition rules from continuum-based models (Guinot, 2002).The emphasis of the next section is on discussing the relationships between partial differential equations and cellular automata, and how to get the transform rules for cellular automata modelling from the physical concepts underlying partial differential equation models.

4.2.1.1 CA and FD for the 1D Diffusion Equation

We first choose the 1D diffusion equation for demonstration purposes:

$$\frac{\partial C}{\partial t} = D\frac{\partial^2 C}{\partial x^2}$$

where: D is the diffusion coefficient; C is the substance concentration.

The Finite Difference (FD) method of discretizing a differential equation on a regular grid with the evolution of states at discrete time steps has similarities with a classical cellular automata approach on the evolution of state variables with a finite number of values on a regular grid at discrete time steps. If the number of states of a cellular automaton is comparable to that of the related finite difference equation, then it can be expected that the results should also be comparable. Using a Forward-Time Central-Space (FTCS) scheme to discrete the 1D diffusion equation, we get

$$C_i^{t+1} = C_i^t + D\frac{\Delta t}{\Delta x^2}(C_{i+1}^t - 2C_i^t + C_{i-1}^t)$$

If the exact solution C_i^t satisfies the equation, then the error value ε_i^t should also meet the discretion equation:

$$\varepsilon_i^{t+1} = \varepsilon_i^t + D\frac{\Delta t}{\Delta x^2}(\varepsilon_{i+1}^t - 2\varepsilon_i^t + \varepsilon_{i-1}^t)$$

Expressing

$$\varepsilon(x) = \sum_{m=1}^{M} A_m e^{ik_m x}$$

with $k_m = \frac{\pi m}{L}$;   m=1,2,....M ; $M = \frac{L}{\Delta x}$

and the amplitude A_m as a function of time t,

$$\varepsilon(x,t) = \sum_{m=1}^{M} e^{at} e^{ik_m x}$$

in which the coefficient a is constant, viz.

$$\varepsilon_m(x,t) = e^{at} e^{ik_m x}$$

Then

$$\varepsilon_i^t = e^{at} e^{ik_m x}$$

$$\varepsilon_i^{t+1} = e^{a(t+\Delta t)} e^{ik_m x}$$

$$\varepsilon_{i+1}^t = e^{at} e^{ik_m(x+\Delta x)}$$

$$\varepsilon_{i-1}^t = e^{at} e^{ik_m(x-\Delta x)}$$

which leads to

$$\varepsilon_i^{t+1} = \varepsilon_i^t + D\frac{\Delta t}{\Delta x^2}(\varepsilon_{i+1}^t - 2\varepsilon_i^t + \varepsilon_{i-1}^t)$$

so that

$$e^{a\Delta t} = 1 + D\frac{\Delta t}{\Delta x^2}(e^{ik_m\Delta x} + e^{-ik_m\Delta x} - 2)$$

Since

$$\cos(k_m\Delta x) = \frac{e^{ik_m\Delta x} + e^{-ik_m\Delta x}}{2}$$

$$\sin^2\frac{k_m\Delta x}{2} = \frac{1 - \cos(k_m\Delta x)}{2}$$

we obtain

$$e^{a\Delta t} = 1 - 4D\frac{\Delta t}{\Delta x^2}\sin^2(k_m\Delta x/2)$$

Defining

$$G = \frac{\varepsilon_i^{t+1}}{\varepsilon_i^t}$$

taking into account the stability criterion

$$|G| \leq 1$$

with

$$G = \frac{e^{a(t+\Delta t)}e^{ik_mx}}{e^{at}e^{ik_mx}} = e^{a\Delta t}$$

and

$$G = 1 - 4D\frac{\Delta t}{\Delta x^2}\sin^2(k_m\Delta x/2)$$

$$\left|1 - 4D\frac{\Delta t}{\Delta x^2}\sin^2(k_m\Delta x/2)\right| \leq 1$$

So $0 \leq D\frac{\Delta t}{\Delta x^2} \leq \frac{1}{2}$

If $D\frac{\Delta t}{\Delta x^2} = \frac{1}{2}$

$$C_i^{t+1} = \frac{1}{2}[C_{i+1}^t + C_{i-1}^t] \tag{4-1}$$

This computational stencil is similar to a structured cellular automaton in one dimension with two-neighbouring cells having equal weight. The truncation error can be analysed from the discrete steps of the partial differential equations as follows:

$$C_{i+1}^t = C_i^t + \left(\frac{\partial C}{\partial x}\right)_i^t \Delta x + \frac{1}{2}\left(\frac{\partial^2 C}{\partial x^2}\right)_i^t \Delta x^2 + \frac{1}{3!}\left(\frac{\partial^3 C}{\partial x^3}\right)_i^t \Delta x^3 + \cdots$$

$$C_{i-1}^t = C_i^t - \left(\frac{\partial C}{\partial x}\right)_i^t \Delta x + \frac{1}{2}\left(\frac{\partial^2 C}{\partial x^2}\right)_i^t \Delta x^2 - \frac{1}{3!}\left(\frac{\partial^3 C}{\partial x^3}\right)_i^t \Delta x^3 + \cdots$$

$$C_i^{t+1} = C_i^t + \left(\frac{\partial C}{\partial t}\right)_i^t \Delta t + \frac{1}{2}\left(\frac{\partial^2 C}{\partial t^2}\right)_i^t \Delta t^2 + \frac{1}{3!}\left(\frac{\partial^3 C}{\partial t^3}\right)_i^t \Delta t^3 + \cdots$$

$$C_i^{t+1} - C_i^t = \left(\frac{\partial C}{\partial t}\right)_i^t \Delta t + \frac{1}{2}\left(\frac{\partial^2 C}{\partial t^2}\right)_i^t \Delta t^2 + \frac{1}{3!}\left(\frac{\partial^3 C}{\partial t^3}\right)_i^t \Delta t^3 + \cdots$$

$$\frac{C_i^{t+1} - C_i^t}{\Delta t} = \left(\frac{\partial C}{\partial t}\right)_i^t + \frac{1}{2}\left(\frac{\partial^2 C}{\partial t^2}\right)_i^t \Delta t^1 + \frac{1}{3!}\left(\frac{\partial^3 C}{\partial t^3}\right)_i^t \Delta t^2 + \cdots$$

$$\left(\frac{\partial C}{\partial t}\right)_i^t = \frac{C_i^{t+1} - C_i^t}{\Delta t} - \left(\frac{\partial^2 C}{\partial t^2}\right)_i^t \frac{\Delta t}{2} + \cdots$$

$$C_{i+1}^t + C_{i-1}^t = 2C_i^t + \left(\frac{\partial^2 C}{\partial x^2}\right)_i^t \Delta x^2 + \left(\frac{\partial^4 C}{\partial t^4}\right)_i^t \frac{\Delta x^4}{12} \cdots$$

$$\left(\frac{\partial^2 C}{\partial x^2}\right)_i^t = \frac{C_{i+1}^t - 2C_i^t + C_{i-1}^t}{\Delta x^2} - \left(\frac{\partial^4 C}{\partial t^4}\right)_i^t \frac{\Delta x^2}{12} + \cdots$$

$$\frac{\partial C}{\partial t} - D\frac{\partial^2 C}{\partial x^2} = 0 = \frac{C_i^{t+1} - C_i^t}{\Delta t} - D\frac{C_{i+1}^t - 2C_i^t + C_{i-1}^t}{\Delta x^2} + \left[-\left(\frac{\partial^2 C}{\partial t^2}\right)_i^t \frac{\Delta t}{2} + D\left(\frac{\partial^4 C}{\partial x^4}\right)_i^t \frac{\Delta x^2}{12} + \cdots \right]$$

If we write the discrete diffusion equation as

$$\frac{C_i^{t+1} - C_i^t}{\Delta t} = D\frac{C_{i+1}^t - 2C_i^t + C_{i-1}^t}{\Delta x^2}$$

then the truncation error is

$$-\left(\frac{\partial^2 C}{\partial t^2}\right)_i^t \frac{\Delta t}{2} + D\left(\frac{\partial^4 C}{\partial x^4}\right)_i^t \frac{\Delta x^2}{12} + \cdots \qquad (4\text{-}2)$$

which is $O[\Delta t, (\Delta x)^2]$.

The cellular automata system is by nature a discrete system, so there is no truncation error. But one should realize that the truncation error should always accounted for if we use CA to simulate a partial equation, because the CA rules were deduced from the discrete equation.

4.2.1.2 CA and FD for the 2D Diffusion Equation

2D diffusion equation: $\dfrac{\partial C}{\partial t} = D(\dfrac{\partial^2 C}{\partial x^2} + \dfrac{\partial^2 C}{\partial y^2})$

The truncation error of the 2D diffusion equation was analysed as follows:

$$C_{i,j}^{t+1} = C_{i,j}^t + \left(\frac{\partial C}{\partial t}\right)_{i,j}^t \Delta t + \frac{1}{2}\left(\frac{\partial^2 C}{\partial t^2}\right)_{i,j}^t \Delta t^2 + \frac{1}{3!}\left(\frac{\partial^3 C}{\partial t^3}\right)_{i,j}^t \Delta t^3 + \cdots$$

$$C_{i,j}^{t+1} - C_{i,j}^t = \left(\frac{\partial C}{\partial t}\right)_{i,j}^t \Delta t + \frac{1}{2}\left(\frac{\partial^2 C}{\partial t^2}\right)_{i,j}^t \Delta t^2 + \frac{1}{3!}\left(\frac{\partial^3 C}{\partial t^3}\right)_{i,j}^t \Delta t^3 + \cdots$$

$$\frac{C_{i,j}^{t+1} - C_{i,j}^t}{\Delta t} = \left(\frac{\partial C}{\partial t}\right)_{i,j}^t + \frac{1}{2}\left(\frac{\partial^2 C}{\partial t^2}\right)_{i,j}^t \Delta t^1 + \frac{1}{3!}\left(\frac{\partial^3 C}{\partial t^3}\right)_{i,j}^t \Delta t^2 + \cdots$$

$$\left(\frac{\partial C}{\partial t}\right)_{i,j}^t = \frac{C_{i,j}^{t+1} - C_{i,j}^t}{\Delta t} - \left(\frac{\partial^2 C}{\partial t^2}\right)_{i,j}^t \frac{\Delta t}{2} + \cdots$$

$$C_{i+1,j}^t = C_{i,j}^t + \left(\frac{\partial C}{\partial x}\right)_{i,j}^t \Delta x + \frac{1}{2}\left(\frac{\partial^2 C}{\partial x^2}\right)_{i,j}^t \Delta x^2 + \frac{1}{3!}\left(\frac{\partial^3 C}{\partial x^3}\right)_{i,j}^t \Delta x^3 + \cdots$$

$$C_{i-1,j}^t = C_{i,j}^t - \left(\frac{\partial C}{\partial x}\right)_{i,j}^t \Delta x + \frac{1}{2}\left(\frac{\partial^2 C}{\partial x^2}\right)_{i,j}^t \Delta x^2 - \frac{1}{3!}\left(\frac{\partial^3 C}{\partial x^3}\right)_{i,j}^t \Delta x^3 + \cdots$$

$$C_{i+1,j}^t + C_{i-1,j}^t = 2C_{i,j}^t + \left(\frac{\partial^2 C}{\partial x^2}\right)_{i,j}^t \Delta x^2 + \left(\frac{\partial^4 C}{\partial t^4}\right)_{i,j}^t \frac{\Delta x^4}{12} \cdots$$

$$\left(\frac{\partial^2 C}{\partial x^2}\right)_{i,j}^t = \frac{C_{i+1,j}^t - 2C_{i,j}^t + C_{i-1,j}^t}{\Delta x^2} - \left(\frac{\partial^4 C}{\partial x^4}\right)_{i,j}^t \frac{\Delta x^2}{12} + \cdots$$

In the same way

$$\left(\frac{\partial^2 C}{\partial y^2}\right)_{i,j}^t = \frac{C_{i,j+1}^t - 2C_{i,j}^t + C_{i,j-1}^t}{\Delta y^2} - \left(\frac{\partial^4 C}{\partial y^4}\right)_{i,j}^t \frac{\Delta y^2}{12} + \cdots$$

$$\frac{\partial C}{\partial t} - D\frac{\partial^2 C}{\partial x^2} - D\frac{\partial^2 C}{\partial y^2} = 0$$

$$= \frac{C_{i,j}^{t+1} - C_{i,j}^t}{\Delta t} - D\frac{C_{i+1,j}^t - 2C_{i,j}^t + C_{i-1,j}^t}{\Delta x^2} - D\frac{C_{i,j+1}^t - 2C_{i,j}^t + C_{i,j-1}^t}{\Delta y^2}$$

$$+ \left[-\left(\frac{\partial^2 C}{\partial t^2}\right)_i^t \frac{\Delta t}{2} + D\left(\frac{\partial^4 C}{\partial x^4}\right)_i^t \frac{\Delta x^2}{12} + D\left(\frac{\partial^4 C}{\partial y^4}\right)_{i,j}^t \frac{\Delta y^2}{12} + \cdots \right]$$

which implies that the truncation error is of the form

$O[\Delta t, (\Delta x)^2, (\Delta y)^2]$

For stability analysis of equation

$$\frac{\partial C}{\partial t} = D(\frac{\partial^2 C}{\partial x^2} + \frac{\partial^2 C}{\partial y^2})$$

We use the general discrete scheme as:

$$\frac{C_{i,j}^{t+1} - C_{i,j}^t}{\Delta t} = D\left[(1-\theta)\left[\frac{\delta_x^2 C_{i,j}^t}{\Delta x^2} + \frac{\delta_y^2 C_{i,j}^t}{\Delta y^2}\right] + \theta\left[\frac{\delta_x^2 C_{i,j}^{t+1}}{\Delta x^2} + \frac{\delta_y^2 C_{i,j}^{t+1}}{\Delta y^2}\right]\right]$$

Let

$$C_{i,j}^t = A^t e^{i(k_x x_i + k_y y_j)}$$

$$G = \frac{A^{n+1}}{A^n} = \frac{1 - 4(1-\theta)\left\{\sigma_x sin^2\left[\frac{k_x \Delta x}{2}\right] + \sigma_y sin^2\left[\frac{k_y \Delta y}{2}\right]\right\}}{1 + 4\theta\left\{\sigma_x sin^2\left[\frac{k_x \Delta x}{2}\right] + \sigma_y sin^2\left[\frac{k_y \Delta y}{2}\right]\right\}}$$

Set

$$B = \sigma_x sin^2\left[\frac{k_x \Delta x}{2}\right] + \sigma_y sin^2\left[\frac{k_y \Delta y}{2}\right]$$

$$G = \frac{1 - 4(1-\theta)B}{1 + 4\theta \cdot B}$$

When $\theta \geq \frac{1}{2}$ set $\theta = \frac{1}{2} + \vartheta$ where $0 \leq \vartheta \leq \frac{1}{2}$

So $G = \frac{1 - 4B\vartheta - 2B}{1 + 4B\vartheta + 2B}$

$$|G| = \left|\frac{1 + 4B\vartheta - 2B}{1 + 4B\vartheta + 2B}\right|$$

$$|1 + 4B\vartheta - 2B| \leq |1 + 4B\vartheta| + |2B| = 1 + 4B\vartheta + 2B = |1 + 4B\vartheta + 2B|$$

So when $\theta \geq \frac{1}{2}$, the discrete form is always stable.

When $\theta \leq \frac{1}{2}$ set $\theta = \frac{1}{2} - \vartheta$ where $0 \leq \vartheta \leq \frac{1}{2}$

$$G = \frac{1 - 4B\vartheta - 2B}{1 - 4B\vartheta + 2B}$$

$$|G| = \left|\frac{1 - 4B\vartheta - 2B}{1 - 4B\vartheta + 2B}\right|$$

If $1 - 4B\vartheta \geq 0, |G| \leq 1$

If $1 - 4B\vartheta < 0, |G| > 1$

Thus the stable condition is

$$B \leq \frac{1}{4\vartheta} = \frac{1}{2(1 - 2\theta)}$$

$$\Leftrightarrow \sigma_x + \sigma_y \leq \frac{1}{2(1 - 2\theta)}$$

$$\Leftrightarrow D \cdot \Delta t \left[\frac{1}{\Delta x^2} + \frac{1}{\Delta y^2}\right] \leq \frac{1}{2(1 - 2\theta)}$$

When $\theta = 0$ (the FTCS scheme) the stability condition reads:

$$D\Delta t \left[\frac{1}{\Delta x^2} + \frac{1}{\Delta y^2}\right] \leq \frac{1}{2}$$

In the particular case with $\Delta x = \Delta y$ this becomes

$$D\frac{\Delta t}{\Delta x^2} \leq \frac{1}{4} \tag{4-3}$$

If $\quad D\frac{\Delta t}{\Delta x^2} = \frac{1}{4}\quad$ then $\quad C_{i,j}^{t+1} = \frac{1}{4}\left(C_{i+1,j}^t + C_{i-1,j}^t + C_{i,j+1}^t + C_{i,j-1}^t\right)$ $\tag{4-4}$

If $\quad D\frac{\Delta t}{\Delta x^2} = \frac{1}{5}\quad$ then $\quad C_{i,j}^{t+1} = \frac{1}{5}\left(C_{i+1,j}^t + C_{i-1,j}^t + C_{i,j}^t + C_{i,j-1}^t + C_{i,j}^t\right)$ $\tag{4-5}$

The second expression is similar to a CA-stencil with the "Parity Rule" (Section 2.4.2). If the state variables are discrete, they are exactly the finite-state cellular automata. However, if the state variables are continuous, they become continuous-valued cellular automata, which are studied by Ostrov and Rucker (1996)

In some sense, continuous-valued cellular automata are similar to finite difference methods, but there are some subtle differences and advantages of cellular automata over finite difference simulations due to e.g. CA's parallel computing capabilities for searching large phase space, as observed by (Rucker et al., 2003)

4.2.1.3   UCA and Finite volume methods (FVM) for 2D diffusion equation

For the 2D diffusion equation:

$$\frac{\partial C}{\partial t} = D\left(\frac{\partial^2 C}{\partial x^2} + \frac{\partial^2 C}{\partial y^2}\right)$$

the discretization based on the triangular mesh using the FVM method becomes

$$\frac{C_i^{t+1} - C_i^t}{\Delta t} V_i = \sum F_{ij} = \sum_{j=1}^{3} \frac{L_{ij}}{d_{ij}} D_{ij} \, (C_j^t - C_i^t)$$

$$\frac{V_i}{\Delta t} C_i^{t+1} = \left(\frac{V_i}{\Delta t} - \sum_{j=1}^{3} \frac{L_{ij}}{d_{ij}} D_{ij}\right) C_i^t + \sum_{j=1}^{3} \frac{L_{ij}}{d_{ij}} D_{ij} \, C_j^t$$

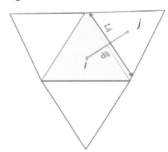

Where:

C_i^t ---the concentration of element i at time step t

F_{ij} ---the normal flux of diffusion

V_i --- the area of the triangular elements i

L_{ij}--- the edge length of edge ij

d_{ij}--- the normal projection of the distance between the centre of i and j

D_{ij}--- the diffusion coefficient between element i and j

For stability reasons, it is required that

$$\sum_{j=1}^{3} \frac{L_{ij}}{d_{ij}} D_{ij} \leq \frac{V_i}{\Delta t}$$

which implies that every element i and its neighbouring element j should satisfy the stable conditions:

$$\frac{L_{ij}}{d_{ij}} D_{ij} \leq \frac{1}{3} \frac{V_i}{\Delta t}$$

$$\frac{L_{ij}}{d_{ij}} D_{ij} \frac{\Delta t}{V_i} \leq \frac{1}{3}$$

$$D_{ij} \frac{\Delta t}{d_{ij}} \frac{L_{ij}}{V_i} \leq \frac{1}{3}$$

$$\Delta t \leq \frac{1}{3} \frac{d_{ij}}{D_{ij}} \frac{V_i}{L_{ij}}$$

Because

$$V = \frac{1}{2} L h$$

In which h is the height of the triangular element

Then we have

$$\Delta t \leq \frac{1}{6} \frac{h_{ij}}{D_{ij}} d_{ij} \tag{4-6}$$

If $h_{ij} \geq 2D_{ij}$

then we obtain

$$\Delta t \leq \frac{1}{3} d_{ij} \tag{4-7}$$

This result shows that Δt is related to the value d_{ij} which is the distance between neighbouring elements. Using the parameter d_{min} which is the minimum value of the d_{ij} among all the triangular elements, Equation (4-7) could be simplified as:

$$\Delta t \leq \frac{1}{3} d_{min} \tag{4-8}$$

Since there are no Δt concepts in cellular automata schemes, for every execution step the CA rules were based only on cell neighbour radius. During the numerical experiments in this thesis (Section 4.3), the numerical value $\frac{1}{3}d_{min}$ was used instead of Δt in order to satisfy the stability condition

$$C_i^{t+1} = \left(1 - \frac{\Delta t}{V_i}\sum_{j=1}^{3}\frac{L_{ij}}{d_{ij}}D_{ij}\right)C_i^t + \frac{\Delta t}{V_i}\sum_{j=1}^{3}\frac{L_{ij}}{d_{ij}}D_{ij}\,C_j^t$$

$$C_i^{t+1} = C_i^t + \frac{\Delta t}{V_i}\sum_{j=1}^{3}\frac{L_{ij}}{d_{ij}}D_{ij}(\,C_j^t - C_i^t\,)$$

$$C_i^{t+1} = C_i^t + \frac{1}{3}\frac{d_{min}}{V_i}\sum_{j=1}^{3}\frac{L_{ij}}{d_{ij}}D_{ij}(\,C_j^t - C_i^t\,) \qquad\qquad (4\text{-}9)$$

4.2.1.4 CA and its relations to specific differential equations

(1) Wave equation

For the 1D linear wave equation,

$$\frac{\partial^2 u}{\partial t^2} = c^2\,\frac{\partial^2 u}{\partial x^2}$$

where c is the wave speed, the discrete version based on a central difference scheme reads

$$\frac{u_i^{t+1} - 2u_i^t + u_i^{t-1}}{\Delta t^2} = c^2\,\frac{u_{i+1}^t - 2u_i^t + u_{i-1}^t}{\Delta x^2}$$

$$u_i^{t+1} + u_i^{t-1} = 2u_i^t + \frac{c^2\Delta t^2}{\Delta x^2}(u_{i+1}^t - 2u_i^t + u_{i-1}^t)$$

$$u_i^{t+1} + u_i^{t-1} = 2u_i^t(1 - \frac{c^2\Delta t^2}{\Delta x^2}) + \frac{c^2\Delta t^2}{\Delta x^2}(u_{i+1}^t + u_{i-1}^t)$$

which can be written as:

$$u_i^{t+1} + u_i^{t-1} = g(u^t) \qquad\qquad (4\text{-}10)$$

where

$$g(u^t) = 2u_i^t(1 - \frac{c^2\Delta t^2}{\Delta x^2}) + \frac{c^2\Delta t^2}{\Delta x^2}(u_{i+1}^t + u_{i-1}^t)$$

This is a reversible function under certain conditions because the wave equation is invariant under transformation from t to -t. (Yang and Young, 2006).

(2) Burgers' equation with white noise

The Burgers equation reads

$$\frac{\partial u}{\partial t} + u\frac{\partial u}{\partial x} = \frac{\partial^2 u}{\partial x^2} + \nabla u$$

By adding Gaussian white noise the equation can be rewritten as

$$\frac{\partial u}{\partial t} + u\frac{\partial u}{\partial x} + \xi = \frac{\partial^2 u}{\partial x^2} + \eta$$

where ξ and η are uncorrelated in space and time.

By introducing the variable

$$v_i^t = c\exp(\Delta x * u_i^t)$$

$$\phi_i^t = \beta\ln(v_i^t)$$

$$\alpha = \frac{\Delta t}{\Delta x^2}$$

$$\exp(-\frac{A}{\beta}) = \frac{1-2\alpha}{c\alpha}$$

$$c^2 = \exp(\frac{B}{\beta})$$

$$\xi = \exp(\phi)$$

$$\eta = \exp(\psi)$$

then in the limit when β tends to zero, the CA rule becomes

$$\phi_i^{t+1} = \phi_{i-1}^t + \max\left[0, \phi_i^t - A, \phi_i^t + \phi_{i+1}^t - B, \psi_i^t - \phi_{i-1}^t\right] - \max\left[0, \phi_{i-1}^t - A, \phi_i^t + \phi_{i-1}^t - B, \phi_i^t - \phi_{i-1}^t\right]$$

$$\lim_{\phi\to 0}\varepsilon\ln(e^{\frac{A}{\varepsilon}} + e^{\frac{B}{\varepsilon}} + ...) = \max[A, B,...]$$

(4-11)

This equation demonstrates that a generalized probabilistic cellular automaton is in essence related to the Burgers equation. Because the Burgers equation was used to present a shock wave without noise, the addition of some noise can be simulated by probabilistic cellular automata which can capture the disorganization of a shock wave (Yang and Young, 2006).

4.2.2   General comparison of CA and PDE

(Wolfram, 1985; Wolfram, 2002) has shown that Cellular Automata and differential equations are related in case of continuum models for physical, chemical and biological processes. Continuum models have advantages such as high accuracy and conservation laws, but mathematical analyses are usually very difficult and analytical solutions do not always exist. There are vast literatures concerning numerical algorithms and numerical solutions of partial differential equations. These include Finite Difference methods which work very well for many problems but have some disadvantages in dealing with irregular geometries; Finite Volume and Finite Element methods which can deal with irregular geometries. Partial differential equations are suitable for systems with only a small degree of freedom and evolutions of system variables in a continuous and smooth manner.

On the other hand, Cellular Automata are often considered to be an alternative mathematical approach. There are many references on the relationship between the two, for example, cellular automata have been used to simulate fluid flows and to solve the Navier-Stokes equations by a Lattice-gas approach. However, Lattice-gas CA is an idealized system where space and time are discrete values. It is mostly restricted to very small scales. For real applications of Navier-Stokes, Lattice-gas CA is not practical (Chen, 2004).

Cellular Automata have the advantage that they can represent discrete entities directly. CA can reproduce the emergent properties of behaviours, and they have large degrees of freedom. Compared with PDE-based models, another unique characteristic of CA is that CA could generate dynamic patterns that are self-reproducing.

CA uses vast numbers of cells or nodes that can take a number of states which always cause a mass of computational loads. And sometimes CA-based modelling is not as accurate and conservative as PDE models. However, CA has universal computability and the nature of parallel implementation; it is powerful and is gradually becoming an essential part of numerical computation.

The following Table 4-2 summarizes the differences between cellular automata based and PDE based modelling:

Table 4-2 differences between CA based modelling and PDE based modelling

	PDE model	CA model
Degree of freedom	Small	Big
Variable State	Continuum	Discrete
Variable Value	Spatial Mean value	Individual (could be)
Emergent properties	No	Yes (could be)
Self-reproducing	No	Yes
Accurate / Conservative	Yes	No (could be)
Parallel implementation	No	Yes

4.3 Effects of cell size in Unstructured Cellular Automata

Ideally, any simulation result should be grid independent; the result should be stable even when calculated on different computational grids. But CA modelling considers one cell as a participant element, and executes the same rules synchronously for every element. If we use small grids instead of large grids and run the same rules, the result shows different patterns since the smaller elements cause more executing steps among the calculating area. Especially for UCA modelling which is based on unstructured meshes, the size of every element is different. The size effect should be an important factor in the UCA rules.

In this research, in order to eliminate the effects of cell size, the parameter d_{min} was used. Firstly, the distance d_{ij} between every central element was calculated, and the parameter d_{min} is the minimum value among those distances. Take the diffusion problem for instance. The relation between the PDE equation and CA rules based on the unstructured triangular meshes was deduced in Equation (4-4):

$$C_i^{t+1} = C_i^t + \frac{1}{3}\frac{d_{min}}{V_i}\sum_{j=1}^{3}\frac{L_{ij}}{d_{ij}}D_{ij}(\,C_j^t - C_i^t\,)$$

This equation will be used as the evolution rules for UCA to analyse the effects of variable meshes. The diffusion coefficient is this equation will be set as a constant, the non-uniform diffusion coefficient were analyzed in the next Chapter.

In this research, numerical experiments were carried out based on the diffusion problem. The results were compared under two groups of meshes. The first mesh set is the original mesh with local refinement; the second type of meshes is the rough mesh compared with the global refined meshes. The results show that the minimum distance between the cells can serve as a characteristic parameter for unstructured cellular automata.

4.3.1   Original meshes & Locally refined meshes

It was supposed that we should get the same results from two sets of meshes: one is the original, and the other is the mesh with local refinement. Fig 4-2 shows the original mesh on a circular area, while Fig 4-3 gives the local refined meshes based on the Fig.4-2.

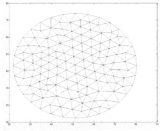

Fig. 4-2 Original Meshes

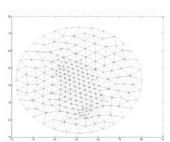

Fig. 4-3 Locally refined Mesh

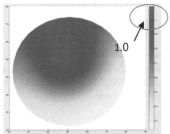

Fig. 4-4 Diffusion on Original Meshes

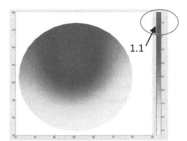

Fig. 4-5 Diffusion on refined Meshes

In this case, after the local refinement, the d_{min} equals 0.8605. Fig. 4-4 and Fig.4-5 are snapshots at 900 steps from the experiments. In these two experiments, d_{min}=0.8605 was adopted. The results show similar diffusion values (maximum concentrations 1.0 compared with 1.1) and diffusion patterns, which means under this rule and parameter, UCA modelling can be cell independent. After running several experiments, it was found that if the original meshes use the same d_{min} as the locally refined meshes, the results of simulation from two sets of meshes are comparable. The diffusion patterns and concentration values are both similar, and the difference between them decreased with the time steps.

4.3.2   Rough meshes & Globally refined meshes

In this part, the experiments are executed separately on the rough mesh and global refined mesh.

Fig. 4-6 illustrates the rough mesh, while the mesh shown in Fig.4-7 is a globally refined mesh generated based on the rough mesh.

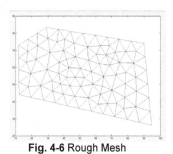

Fig. 4-6 Rough Mesh

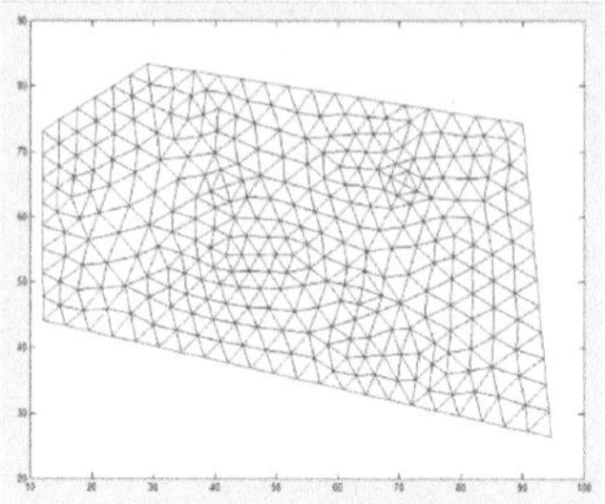

Fig. 4-7 Globally refined Mesh

Using separate d_{min}

For these sets of meshes, we keep their own d_{min} separately in the modelling. The d_{min} which belongs to the rough meshes (Fig.4-6) equals 2.5851, while the d_{min} of the fine mesh (Fig.4-7) is reduced to 1.066 after global refining. If we calculate the ratio between these two d_{min} values, the ratio is around 2.4 times.

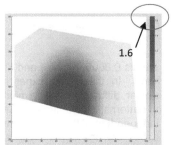

Fig. 4-8 Rough Meshes in 250 steps

Fig. 4-9 Refined Meshes in 600 steps

Running the models and making a snapshot after 250 steps (Fig. 4-8) and a snapshot at 600 steps (Fig. 4-9) separately, it can be seen that similar diffusion patterns and comparable concentration values appear. The proportion approximates the ratio value from d_{min}. For example, Fig.4-8 / Fig.4-9 = 600steps / 250steps = 2.4 (which was mentioned in last paragraph).

Using same d_{min}

In Fig 4-6 and Fig 4-7 we change the parameter to d_{min}=1.066, which is the minimum characteristic distance belonging to the globally refined mesh. With this parameter and executing the model for 800 steps, the results are shown in Fig. 4-10 and Fig. 4-11.

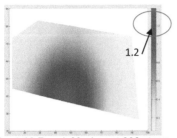

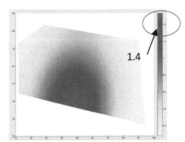

Fig. 4-10 Rough Meshes at 800 steps   **Fig.4-11** Refined Meshes at 800 steps

The results in Fig. 4-10 and Fig. 4-11 show similar diffusion patterns. From the graphs it can be seen that using the same characteristic parameter, UCA modelling under different meshes leads to almost the same diffusion process.

4.3.3   Analysis and discussion

In this research, the minimum distance d_{min} was adopted as a characteristic parameter in unstructured cellular automata modelling. The results show that in most cases, the performance of UCA was comparable even based on different sets of meshes. Generally speaking, the results obtained similar concentration values and diffusion patterns at the same time step if the same d_{min} was adopted. On the first set of meshes, we used the d_{min} which belongs to the locally refined meshes, and in the case of the second type of meshes, the parameter d_{min} was calculated from globally refined meshes. With the same parameter, the results from rough meshes and globally refined meshes are very similar. Meanwhile, if the d_{min} from rough meshes and globally refined meshes are used separately, the results are still comparable but proportional to the time scale, which is related to the ratio of d_{min}.

Chapter 5

Unstructured cellular automata for spatial dynamic ecological modelling

Ecosystem changes continuously in space and time, thus predictions are important for people to accommodate these changes and take proper actions. Models are broadly used and proven to be effective to achieve the objective.

Ratze (Ratze et al., 2007) summarized three common numerical paradigms which were used to represent spatial pattern dynamics: (i) Physical paradigm, based on differential equations, either Ordinary Differential Equations (ODEs) or Partial Differential Equations (PDEs); (ii) Discrete paradigm, including Cellular Automata (CA) and Discrete Event Specification systems (DEVS); (iii) Agent based paradigm, such as individual-based models (IBM) and Multi-Agent Systems (MAS). (Li, 2009)

Although partial differential equations may naturally fit different situations and have been widely used at present, they can be mathematically complex thus difficult to solve. Meanwhile, most of the current ecological models are capable of simulating the conventional system; however, they are not sufficient when describing behaviors of systems under novel conditions (Evans, M. R., 2012). Cellular Automata employ very simple mathematical rules, but has the advantages to describe complex dynamics. This chapter demonstrates the capability of Unstructured Cellular Automata (UCA) in ecological modeling through three applications, including the prey-predator system; algae blooms dynamics; and water quality changes of spiked outlet in Hong Kong.

5.1 UCA for prey-predator model

Prey-predator dynamic is a classical problem of ecosystem evolution. Vito Volterra, a mathematical physicist, first used simplified differential equations to simulate the dynamics of prey and predatory fish populations in the Adriatic in 1926 (Volterra, 1926; Scholarpedia, 1(10):1563.). On one hand, this kind of models has the advantage of mathematical tractability; and most of which can be solved analytically to give precise results. But, on the other hand, it seems obviously that the models can never reflect any particular system accurately since they do not consider realistic conditions.

A Cellular Automaton has been applied for the Prey-Predator models, which can be found in (Chen, 2003; Cattaneo et al., 2006; Arashiro and Tome, 2007; Farina and Dennunzio, 2008). However, classical cellular automata are built on the structured grids, where every element (grid) indicates one prey, one predator, or an empty state. The state of the predator is determined by the defined neighbors. The predator itself can't move to another grid. In classical CA-based prey-predator models, each grid is considered as an individual. But in reality, the distribution and the group number are more important than individuals. In addition, predators' abilities to search for food should be taken into account.

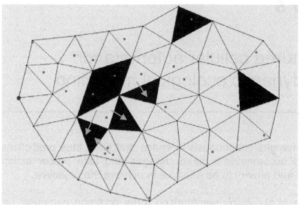

Fig. 5-1 Schematic diagram of animals' movements

Fig. 5-1 shows the schematic diagram of the proposed model, where the dark areas are covered by enough grass (food) and the red points represent individual animals. The yellow arrows indicate the possible routes along which the predators can search for food.

The unstructured grid used in this model is spatially non-uniform. If the density of the predators is relatively high, or the grass distribution is very complicated, the girds can be refined to mimic the reality.

In order to simulate the movements of the vegetarians, optimization techniques are adopted. Assume the vegetarians have the intelligence to search food for survival. If there is enough grass in a grid for vegetarians to feed on, they will stay within the grid; otherwise, some of them will move to neighboring elements (if grass is still redundant in the neighboring elements). Mostly, the vegetarians have three options (neighboring elements) to choose; but in some special locations (such as coroners), the vegetarians have only one direction to move. If grass is limited globally, the searching routes are optimized in order to obtain maximum possible survivors.

The growth and the spread of grass have been taken into account. The reproduction rate of the vegetarians (represented by sheep vegetarian) is also considered.

The initial conditions:
(1) grass density in each element—dgrass(i);
(2) the number of vegetarians in each element—nsheep(i).

......

Some parameters:
(1) the amount of grass a sheep needs within Δt —msheep_need;
(2) reproduction rate of sheep—k1
(3) growth rate of grass—k2
(4) spread rate of grass—k3

......

Some variables:

(1) the maximum number of sheep in the element:

maxsheep(i)=dgrass(i)*area(i)/msheep_need;

(2) the number of unknown variables at each time step—Nunknown
(3) the coefficient array of linear programming problems— A(i, Nunknown)
(4) Total time for running — T
(5) Total elements — Etotal

The procedure can be summarized as following (Fig. 5-2):

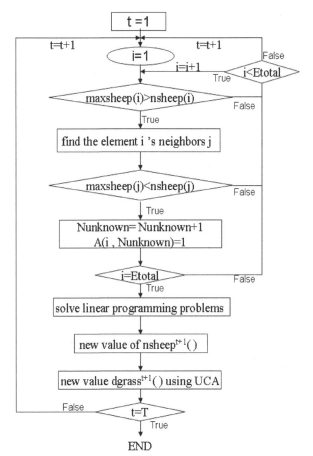

Fig. 5-2 The calculate process of the grass-vegetarian model

Snapshot and population dynamics of the developed grass-vegetarian model are shown in Fig. 5-3 and Fig. 5-4. The green grids stand for the areas covered by grass. The white grids represent the areas with less grass. The small black circles represent the movements' routes of vegetarians.

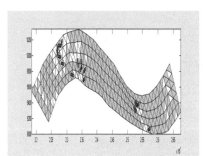

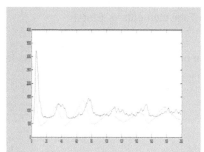

Fig. 5-3 The snapshot at 20 steps after running of the grass- vegetarian model

Fig.5-4 The population dynamics of the grass- vegetarian model

5.2 UCA for algae bloom model

The performance of algal bloom models is restricted by spatial heterogeneities and local interactions, which can be simulated by a cellular automata paradigm. The classical cellular automata paradigm is based on structured grids with predefined interactions between species at a local level. Unstructured meshes can provide the flexibility, which the structured grids are lacking, to focus on resolution on features of interest and avoid the unfortunate staircase representation of boundaries.

In this section, cellular automata model was developed based on the geometry of the North Sea. Both classical CA and unstructured CA were applied and the results were compared.

There are 72554 triangular elements and 35733 vertices in the unstructured mesh, and 17560 curvilinear elements and 18192 vertices in the structured mesh. The curvilinear mesh has fewer grid vertices, at the cost of losing some geometric configuration.

When the initial spatial distribution is "randomly mixed", the two kinds of meshes turn out to have the same regularities: the dynamic competition reaches to an equilibrium state, where the population of the three colors roughly stabilizes around their average. (See Figure 5-5 & Figure 5-6)

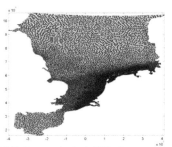

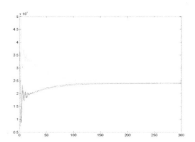

Fig. 5-5 A snapshot at 300 steps under the randomly mixed initial condition based on unstructured meshes (Red10%, Green40%, White 50%)

Fig. 5-6 The population dynamics of randomly mixed initial distribution based on unstructured meshes (Red 10%, Green 40%, White 50%)

The advantage of the Unstructured Cellular Automata was clearly demonstrated when the initial spatial distribution is "one isolated in middle". The Unstructured Cellular Automata spread homogeneously in space (see Figure 5-7); while the curvilinear-based structured Cellular automata are obviously grid dependent (see Figure 5-8).

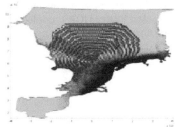

Fig. 5-7 (a) Initial distributed with one color isolated in the middle based on unstructured meshes (Red 2%, Green 49%, White 49%)

Fig. 5-7 (b) Snapshot at 60 steps with one color isolated in the middle in initial based on unstructured meshes (Red2% White49% Green49%)

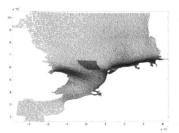

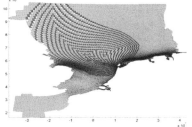

Fig. 5-8 (a) Initial distributed with one color isolated in the middle based on curvilinear meshes (Red2% Green49% White49%)

Fig. 5-8 (b) Snapshot at 50 steps with one color isolated in the middle based on curvilinear meshes (Red2% White49% Green49%)

The developed Unstructured Cellular Automata provide the flexibility for isotropic system, which is missing in classic Cellular Automata, and they also demonstrates their capabilities in ecosystem modeling. The present work focused on the methodology of Unstructured Cellular Automata. More practical studies are to be given to verify and improve the unstructured cellular automata paradigm.

5.3 Spatial water quality model for spiked pollution

5.3.1 Non-uniform diffusion water quality model

It is generally believed that the water quality problem could be simulated by physical transport equations. (Postma et al., 1998; Stelling, 1984.) But for spiked-shape problem, some coefficients vary temporally and spatially. Besides, several factors could not be taken into account by using the physical equations. Therefore, it is difficult to simulate this kind of problem by using the transport equations. An alternative method is the Cellular Automata paradigm, which is proved to be a useful method for simulating non-smooth system.

To simulate the spiked pollution loading phenomena, different diffusion coefficients were tested in the UCA modelling in this research. For x direction the diffusion was set as normal distribution; and in y direction, several numerical experiments were carried out.

Fig 5-9 (a) shows the distribution of diffusion coefficient in x direction, where the normal function was adopt as $Ex \sim N(0, 0.42)$
Fig.5-9 (b) shows the setting of diffusion coefficient in y direction based on the function: $Ey = y/(y+1)$

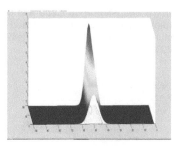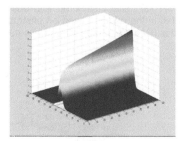

Fig. 5-9 (a) Distribution of diffusion coefficients in x direction (3D view)

(b) Distribution of diffusion coefficients in y direction (3D view)

More coefficient distributions in y direction were tested, as shown in the coefficient curves in Fig. 5-10 together with their corresponding functions. They included the constant coefficients (e.g. $Ey=5$), and the functional settings that gradually increased or decreased along the y direction from the entries of pollution sources on the boundary.

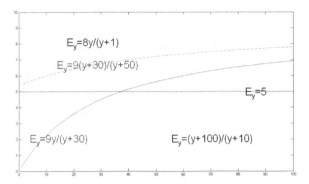

Fig. 5-10 Numerical experiments on diffusion coefficient setting in y direction

With various coefficient settings in 2D space, the simulated result shows different diffusion processes and patterns:

(1)   Ey=5

If a constant coefficient was adopted, compared with other coefficients, the constant value should be set within a small threshold value. For instance, when the maximum of the other coefficients is close to 8, the constant coefficient could not exceed 5. Otherwise, the UCA modelling became unstable.

(2)   Ey=8y/(y+1)

The curve of experiment (2) is above the other curves, which means the value of this setting is higher than the others, and it can cause the quickest diffusion process among all the experiments.

(3)   Ey=9y/(y+30)
(4)   Ey=9(y+30)/(y+50)
(5)   Ey=(y+100)/(y+10)

Experiment (3) diffused very slowly at the beginning. Compared with experiment (4), the diffusion region of experiment (3) is much smaller when they decrease to the same concentration range. To reach the same concentration value, Experiment (5) needs the longest time, but its diffusion area is larger than experiment (4).

After knowing the effect of diffusion coefficients in UCA, some modelling exercises were conducted. By setting the initial concentrate on the bottom of a round boundary, the result from isotropic diffusion coefficient shows the round diffusion patterns (Fig.5-11). It is not suitable for simulating the spiked pattern problem. By using suitable anisotropic and non-uniform coefficient, the spiked patterns could be well captured (Fig.5-12)

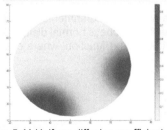

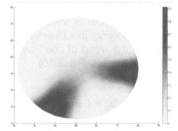

Fig. 5-11 Uniform diffusion coefficient   **Fig. 5-12** Non-uniform diffusion coefficient

5.3.2   Application for study case (Numerical experiment)

In this research, the Victoria harbor was taken as the study case (Fig. 5-13) where the yellow line is the geometric boundary.

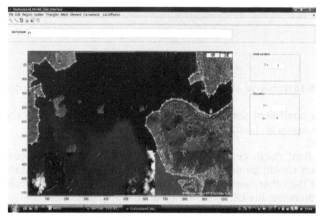

Fig. 5-13 The map of Victoria harbor

The primary meshes were generated using Delaunay triangulation technique (Fig. 5-14). To get more details about the diffusion process, the meshes were refined at the source entries of the pollution (Fig. 5-15). The initial pollution conditions (concentration value and locations) were hypothesized for the simulation experiments.

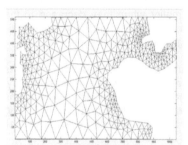   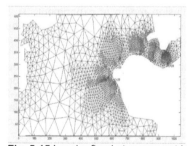

Fig. 5-14 Initial meshes based on geometry   **Fig. 5-15** Local refined at source entries

The diffusion coefficients were set according to the numerical experiments in Section 5.3.1 as the following: in x direction, the diffusion coefficient obeys normal distribution; in y direction, the trend of coefficient is based on the tanh(x) function, where c is an adjustable parameter (Fig. 5-16).

Ex~ $N(0, 0.4^2)$
Ey= - 0.5*tanh(c*x) + 0.5

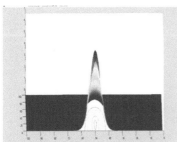

Fig. 5-16 (a) Distribution of diffusion coefficients in x direction

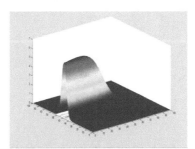

(b) Distribution of diffusion coefficients in y direction

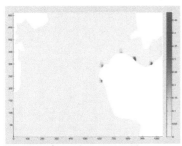

Fig. 5-17 Snapshot at step10

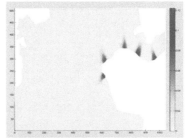

Fig. 5-18 Snapshot at step1000

Fig 5-17 is a snapshot at step 10 which shows the pollutant diffused near the sources region. After 1000 steps, the spiked pattern became obvious, as shown in Fig. 5-18. The simulation results indicated that by using suitable anisotropic and non-uniform diffusion coefficients, the spiked patterns could be captured by UCA modelling. Even for the small-scaled pollution, the pollutant region at source entries could be easily refined and modelled based on the unstructured meshes.

These numerical experiments gave a general idea to apply unstructured cellular automata in the water quality modelling. The developed unstructured cellular automata provide flexibility which is missing in classic cellular automata.

Until this stage, the study focused on the methodology of unstructured cellular automata and its potentials in ecosystem modelling. In the next sections, more realistic simulation studies were carried out to demonstrate the capabilities of unstructured cellular automata paradigm in ecological modelling.

Chapter 6

Spatial evolution of benthonic macroinvertebrate under flow regulation using hybrid modelling

6.1 Introduction

Flow regulations usually imposed great impacts on aquatic ecosystem, such as alteration of hydrological and water quality conditions, disrupt of habitat connectivity, nutrient cycling and sediment transportation, block of fish path and invertebrate movements (Santucci et al., 2005).

In recent years, great efforts have been taken to evaluate the effects of flow variability on aquatic ecosystem (Li et al., 2010; Li et al. 2012; McClain et al. 2014; Ye et al., 2010). Being an important component, benthic macroinvertebrates play a special role in aquatic systems. Therefore, it is valuable to investigate the relationships between flow alterations and macroinvertebrate distributions. Plenty of studies have focused on macroinvertebrates (Chen et al., 2011; Chen et al., 2013; Cortes et al., 2002; Sheldon & Thoms, 2006; Dunbar et al., 2010), and most of these researches have assessed the qualitative impacts. However, with a growing awareness of the value of natural ecosystems, there are strong demands to investigate the impacts of flow regulation quantitatively and to seek for implementable remediation measures (Nagaya et al., 2008; Li et al., 2010).

This research aimed to quantify the spatial changes of macroinvertebrate distribution induced by flow regulation. To simulate the spatial distribution of macroinvertebrates under different flow regulating conditions, a hybrid ecohydraulics model was set up, which couples a water quality module with a macroinvertebrate habitat module. The habitat module was based on an Artificial Neural Networks (ANN) model, which was trained and validated by two-year datasets.

To understand the spatial patterns of macroinvertebrate dynamics, cellular automata techniques were introduced in this study. Two criteria are adopted: patch analysis using cellular automata to generate the patches of macroinvertebrate, and to calculate the areas of the patches; homogeneity characterizing the cluster feature of the entire system (M.-Th. Hutt & R.Neff, 2001).

The realistic scenario together with the hypothesis scenario were implemented and compared.

6.1.1 Description of study area

Lijiang river basin (Fig. 6-1), a world famous scenic spot with natural karst landscape, is located in the Southwest of China. It originates from the Mao'er Mountain and flows from north to south crossing through Guilin, Yangshuo and Pingle cities. (Chen, et.al., 2013)

The main stream of the Lijiang River is 214 km long, and the total catchment area is 12,285 km^2.(Miao, 1997) The Lijiang River is generally divided into three reaches according to the physical features. The upstream extends from Mao'er Mountain to Guilin city, where the substrate is mainly mud and fine sand. The middle stream, down to the Yangshuo County, is characterized as transitional zone with high flow velocity and pebble sediment. The downstream is to the confluence of Guijiang River, where the sediment mainly consists of pebble and gravel. Due to the special karst landscape and the strong seasonality of rainfall, the daily averaged discharges in the Lijiang River vary from 12 m^3/s to 12,000 m^3/s, with an annual average of 120 m^3/s. The recorded minimum discharge was 8 m^3/s, which posed great threats to the local water supply and aquatic ecosystem. Therefore, a series of reservoirs have been or will be constructed in the main stream and the upstream tributaries to overcome this problem. At present, the Qingshitan reservoir is in full operation, and the Chuanjiang reservoir was completed in 2012. The Darongjiang reservoir is under construction.(Chen, et.al., 2013)

With the regulation of the Qingshitan reservoir, the discharge in the dry season was increased from 12 m^3/s to 28 m^3/s. After all the reservoirs are in operation, the discharge in the dry season will raise up to 60 m^3/s. Since the hydro-environmental conditions have been dramatically altered by the operating reservoir and will be further modified, it is important to investigate their effects on the aquatic ecosystem. Previous studies focused on the fish habitat (Li et al., 2010) and riparian vegetation evolution (Ye et al., 2010), while this research concentrated on macroinvertebrates, and particularly using unstructured cellular automata to quantify the spatial-tempo dynamics of macroinvertebrates.

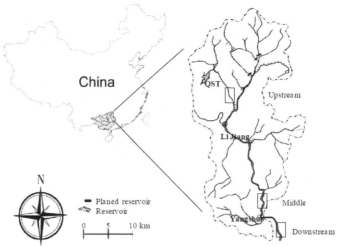

Fig. 6-1 Lijiang River Basin and data collection reaches, QST: Qingsitan Reservoir (Chen et al., 2011)

6.1.2 Data Collection

In total, three reaches (separately located at upstream, middle reach, downstream, See Fig. 6-1 were selected, in which samples for physical, chemical and biological analyses were taken. These samples were examined twice over a 2-year period (autumn of 2009 and summer of 2010), and the database consisted of 300 instances.

At each site, physical and chemical parameters were measured prior to macroinvertebrate sampling. Temperature (℃), conductivity (mS/cm), turbidity (NTU), dissolved oxygen (mg/L), and pH were measured in situ using a portable YSI (YSI 6600). Water depth (cm) was measured with a wading rod and flow velocity (m/s) was measured by a hydrometric propeller near the bed for 60s time interval.

Water samples for chemical parameter analyses were collected at the position about 10-15 cm above the river bed at each microhabitat, and were preserved in 500 ml polyethylene bottle. All the water samples were placed in an ice chest at 4℃, and were analyzed immediately after the samples arrived in the laboratory, which was within 24 h after collection.

The total nitrogen (mg/L) and total phosphorus (mg/L) were determined by UV spectrophotometer. The chemical oxygen demand (COD, mg/L) was analyzed by potassium dichromate colorimetric method.

The grab sample bed material type was monitored visually. Manning coefficient was calculated depending on reliable and consistent evaluations of channel condition and an accurate measurement of the cross-sectional area, hydraulic radius, and slope.

The hydrological data during 2008-2010 were collected from the local hydrological stations, including daily averaged discharge, water level, and water quality parameters. Furthermore, the bathymetry and the flow profiles of the studied river section were measured by the Doppler flow measurement device--River Cat. In total, 100 cross-sections (with an interval of 50 m) were measured and the bathymetry of the entire area was obtained by interpolation.

Macroinvertebrate living in different habitats were sampled using a perterson grab dredger (1/20 m^2). The sampled materials were first put in a plastic basin, then the materials with size below 0.5 mm were eliminated by means of sieving using a metal sieve, and finally the remaining materials were preserved in 100ml plastic bottles with 5% formaldehyde solution. The macroinvertebrate taxa were identified in the laboratory under a stereoscopic dissection microscope (magnification 10~75 times).

In total, there are 300 instances sampled from the three reaches of Lijiang River, including 108 instances in the upstream (Tan Xia), 90 instances in the middle stream (Da Yu), and 102 samples in the downstream (Fu Li).

6.2 Hybrid Model developments

The hybrid model was applied to the compound channel in the middle reach of Lijiang River, where the flow was seriously regulated by the Qingshitan reservoir upstream. The integrated model simulated water temperature, dissolved oxygen, water depth, water velocity and the distribution of *Semisulcospira amurensis* (Chen et al., 2011)

6.2.1   Two-dimensional water quality module

To model the spatial distribution of macroinvertebrate, the study area is divided into 20m*20m meshes, as shown in Fig. 6-2.

Fig. 6-2 Mesh Generation (Middle stream)

Basing on the measured cross-sectional data, the bathymetry of the entire river reaches was obtained through trigonometric interpolation method, as shown in Fig. 6-3.

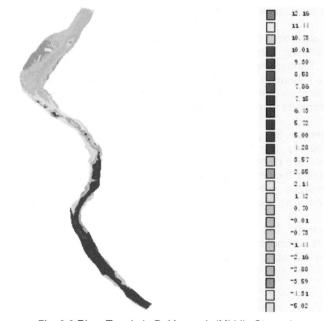

Fig. 6-3 River Terrain in DaYu reach (Middle Stream)

Delft-3D software package was used to simulate the hydrodynamic and water quality processes. In the flow module, daily averaged discharge was used at the upstream boundary and daily water level was used at the downstream boundary. The time step was set to 6 seconds.

There modeled environmental factors, affecting the spatial distribution of macroinvertebrate, included flow regime (water depth and flow velocity) and water quality (water temperature, dissolved oxygen, bed material type). The hydrodynamics was modeled by the two-dimensional shallow water equations (6-1)-(6-3):

$$\frac{\partial H}{\partial t} + \frac{\partial (hu)}{\partial x} + \frac{\partial (hv)}{\partial y} = Q_a \tag{6-1}$$

$$\frac{\partial u}{\partial t} + u\frac{\partial u}{\partial x} + v\frac{\partial u}{\partial y} = -\frac{1}{\rho_0}\frac{\partial p}{\partial x} + fv + v\left(\frac{\partial^2 u}{\partial x^2} + \frac{\partial^2 u}{\partial y^2}\right) + \frac{1}{\rho_0 H}\tau_x \tag{6-2}$$

$$\frac{\partial v}{\partial t} + u\frac{\partial v}{\partial x} + v\frac{\partial v}{\partial y} = -\frac{1}{\rho_0}\frac{\partial p}{\partial y} - fu + v\left(\frac{\partial^2 v}{\partial x^2} + \frac{\partial^2 v}{\partial y^2}\right) + \frac{1}{\rho_0 H}\tau_y \tag{6-3}$$

where, Q_a is discharge or withdrawal (m³/s) , H is water level (m), u, v are velocity in x and y direction (m/s), υ is horizontal eddy viscosity coefficient (m²/s), f is Coriolis parameter, τ_x, τ_y are bottom shear stress. Implicit scheme was applied to solve the equations numerically.

Meanwhile, the water quality was modeled by the two-dimensional advection-diffusion equations (6-4) with source/sink and reaction terms:

$$\frac{\partial c}{\partial t} + u\frac{\partial c}{\partial x} + v\frac{\partial c}{\partial y} = D_x\frac{\partial^2 c}{\partial x^2} + D_y\frac{\partial^2 c}{\partial y^2} + S + f_R(c,t) \tag{6-4}$$

where, c is concentration (mg/L), D_x and D_y is dispersion coefficients (m²/s), S is source or sink term and $f_R(c, t)$ is reaction term.

It is well known that reservoir operation dramatically changed water temperature and dissolved oxygen concentrations that have great impact on river ecosystem. Therefore, water temperature and dissolved oxygen concentrations were taken into consideration in the water quality module, and bed material type is defined as constant value.

The monitored data of environmental factors were grouped into the dry season and the high flow season. The mean value of monitoring data in high flow period was selected for model parameter calibration. Some Parameters were listed in Table 6-1.

Table 6-1 Parameters in Two-dimensional Water Quality module

Parameter	Value	Parameter	Value
C_P (J/(kg ℃)	4.2×10^3	ρ (kg/m³)	1000
Φ_{so} (W/m²)	188.27	C_S (g/m³)	7.30×10^{-3}
C (%)	0.3	$K_{1,20}$(1/d)	0.1
W (m/s)	1.00	i (‰)	1.34
T_S (℃)	32.0	U(m/s)	0.13
T_d(℃)	26.5	H(m)	0.87
T_a(℃)	28.0		

where:

C_p— Specific heat of water

Φ_{so}— the total solar radiation during sunny days

C — the cloud cover ratio

W -- the wind speed at 10m on the water

Ts-- the temperature of the water surface

T_d-- the dew point temperature

T_a-- Air temperature on the water at 2m

ρ -- the density of water

C_s--the saturation concentration of dissolved oxygen under temperature T

$K_{1,20}$ --the BOD5 degradation rate constant (temperature=20)

i-- the river slope

U --the flow velocity

H-- the water depth

CBOD degradation rate takes value from 0.04 1/d to 0.08 1/d. Due to the characteristics of the river in terms of bed materials, meandering shape, and flat bathymetry, Manning roughness coefficient ranges from 0.04 to 0.05.

6.2.2   macroinvertebrate habitat module

The aim of macroinvertebrate habitat module is to predict the possibility of *S. amurensis* presence. Due to the high complexity, nonlinearity and insufficient knowledge to the relation between hydro-environmental conditions and macroinvertebrate presence (Chon et al., 2002, Chen and Mynett, 2006), the Habitat module was developed based on an artificial neural network (ANN). A three layer feed-forward back-propagation neural networks was constructed in order to map the presence of *S. anmurensis* from hydro-environmental parameters. (Chen et al., 2011)

In the macroinvertebrate habitat module, five hydro-environmental parameters were selected as the inputs of ANN model, which included water temperature, dissolved oxygen, water depth, water velocity and manning coefficient (Table 6-2). The absence/presence of *S. amurensis* is taken as the output. Thus the ANN network consisted of 5 inputs and 1 output (Chen et al., 2011).

Table 6-2 Variables and units used in the ANN model

Variables	Units	Min	Max	Mean	SD
Temperature	°C	15.99	34.11	23.44	5.84
Water depth	cm	5.6	686	109.2	96.1
Flow velocity	m/s	0	1.869	0.287	0.339
Dissolved oxygen	mg/L	6.12	12.56	9.79	1.83
Manning coefficient	4 classes (silt =0.033; silt and grass=0.039; cobble=0.042; cobble and grass=0.046)	0.033	0.046	0.042	0.003

SD: standard deviation

The presence and absence of macroinvertebrate are represented by S = 1 and S = 0, respectively. And the continuous values from model outputs are classified into 1 or 0 using a threshold of 0.5 (Dedecker et al., 2004).

$$S = \begin{cases} 1 & a \geq 0.5 \\ 0 & a < 0.5 \end{cases}$$

in which a is the ANN model output value.

6.2.3 Model verification

The distribution of S. amurensis was verified under current flow condition (with discharge equal to 28 m^3). Fig.6-4 and Fig. 6-5 showed the snapshots of four environmental factors (Water temperature, Dissolved oxygen, water velocity and water depth) modeled by the two-dimensional water quality module (described in Section 6.2.1). Using the modeled values of the environmental factors as the input (together with manning coefficient), ANN module simulated the distributions of S. amurensis, as shown in Fig. 6-6.

It can be seen that compared with the filed data of S. amurensis which was collected in September 2009, the hybrid model reproduced the similar distribution of S. amurensis.

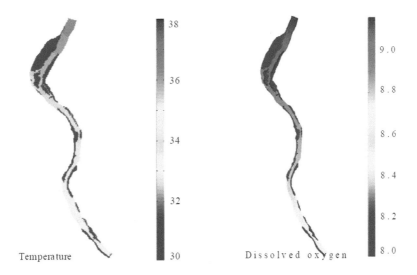

Fig. 6-4 The snapshot of water temperature and dissolved oxygen when discharge is 28 m^3

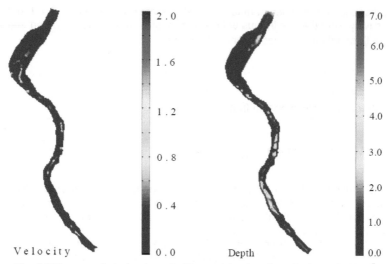

Fig. 6-5 The snapshot of water velocities and depths when discharge is 28 m^3

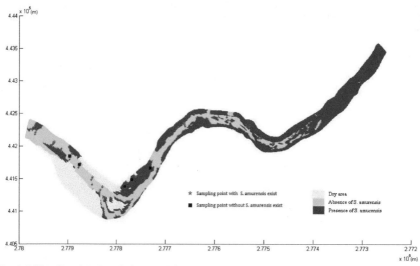

Fig. 6-6 The Simulated and observed distribution patterns of *S. amurensis* under current flow regime (model verification)

6.2.4   Scenario analyses

Under the regulation of reservoir, the discharge will raise from 28 m^3, to 60 m^3, which will change the hydro-environmental conditions dramatically. Figure 6-7 and Figure 6-8 presented the modeled flow velocity, water depths, water temperature and dissolved oxygen for the discharge of 60 m^3.

When the flow is 28 m³/s, the water temperature in the studied area ranges from 30 to 37.68 °C, with an average of 31.93 °C; dissolved oxygen ranges from 8.03~ 9.15 mg/L, with an average of 8.77 mg/L; flow velocity varies between 0.60 m/s and 1.73 m/s, with an average of 0.12 m/s; water depth ranges from 0 cm to 544.9 cm, with an average of 87.2 cm.

When the flow is raised to 60 m³/s, water temperature in the studied area ranges from 30 to 37.19 °C, with an average of 31.91 °C; dissolved oxygen ranges from 8.03 to 9.14 mg/L, with an average of 8.57 mg/L; flow velocity varies between 0 and 1.91 m/s, with an average of 0.18 m / s; water depth ranges from 0 to 671.1 cm, with an average of 148.1 cm.

Table 6-3 Water Environmental factors under two scenarios

Flow	28 m³/s,			60 m³/s		
	Min	Max	Average	Min	Max	Average
Water Temperature (°C)	30	37.68	31.93	30	37.19	31.91
Dissolved Oxygen (mg/L)	8.03	9.15	8.77	8.03	9.14	8.57
Flow Velocity (m/s)	0	1.73	0.12	0	1.91	0.18
Water Depth (cm)	0	544.9	87.2	0	671.1	148.1

From Table 6-3, it can be seen that compared to the current condition of 28 m³, the average water depth and water velocity increased significantly. The average water depth increased with 60.9 cm, leading to the decrease of dry riverbed from 5.12×10^{5} m² to 4.69×10^{5} m², which directly affected the macroinvertebrate distribution. The other two factors, water temperature and dissolved oxygen, were also simulated, but their changes are slight.

After water replenishing, hydro-environmental conditions of the river has a significant difference during the dry season. The riparian area decreased, while the deepwater area increased.

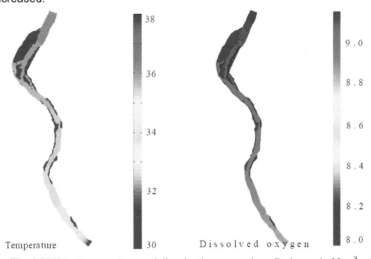

Fig. 6-7 Water temperature and dissolved oxygen when discharge is 60 m³

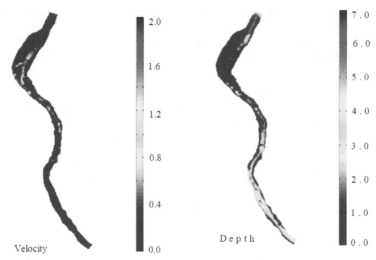

Fig. 6-8 the snapshot of water velocities and depths when discharge is 60 m³

Using the modeled results of the environmental factors as inputs, the macroinvertebrate habitat model can simulate the spatial evolution of benthonic macroinvertebrate under flow regulation.

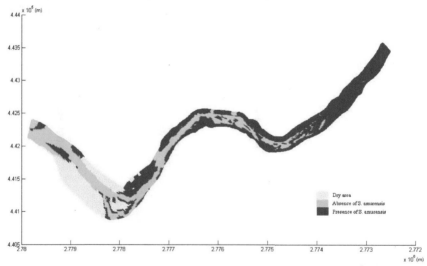

Fig. 6-9 the macroinvertebrate distribution pattern when discharge is 28 m³ (Yellow color: bank area, green color: presence of macroinvertebrate, blue color: absence of macroinvertebrate)

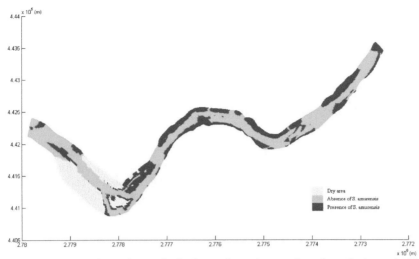

Fig. 6-10 the macroinvertebrate distribution pattern at scenarios when discharge is 60 m³ (Yellow color: bank area, green color: presence of macroinvertebrate, blue color: absence of macroinvertebrate)

The modeled macroinvertebrate habitat distribution of the two scenarios was shown in Fig.6-9 and Fig6-10. When the discharge is 28 m³/s(Fig 6-9), the bank area was 5.12×10⁵ m², the empty area was 8.06×10⁵ m², and the potential habitat area was 11.35×10⁵ m². When the discharge is 60 m³/s (Fig 6-10), the bank area was 4.69×10⁵ m², the empty area was 13.49×10⁵ m², and the potential habitat area was 6.36×10⁵ m²

Table 6-4 Comparison of S. amurensis distribution area
between the current flow regime and designed flow

Discharge (m³/s)	Presence area (m²)	Dry area (m²)	Absence area (m²)
28	11.36 ± 0.013×10⁵	5.12×10⁵	8.04± 0.014×10⁵
60	6.39 ± 0.009×10⁵	4.68×10⁵	13.46± 0.008×10⁵

From Table 6-4, it can been seen that after flow regulation, the S. amurensis habitat area was decreased from 1.35×10⁵ m² to 6.36×10⁵ m², with the reduction percentage of 44%.

The above results showed the changes of habitat area due to flow regulation. However, it is also important to know the effects on spatial distribution and geometry fraction of macroinvertebrate' habitat. The next section will apply unstructured cellular automata method to quantify the spatial patterns of macroinvertebrate habitats.

6.3 Quantify spatial distribution of macroinvertebrate using cellular automata

6.3.1 Patch analysis of macroinvertebrate habitat using cellular automata

In this section, the cellular automata technique was applied to analyze the spatial pattern of macroinvertebrate habitat. (Lin, et al., 2011a) 'Four neighbors' scheme was implemented in the CA paradigm (Figure3-1 (a)).

The patch searching procedure is given in the following:

(1) Firstly, the state (presence/ absence) of each CA element from the habitat module is checked. If the element's state is 'presence', this element is considered as an individual patch at the beginning, and is assigned a non-redundant patch ID number as the initial condition in the CA modeling.
(2) Secondly, run the CA model synchronously. If any of the neighbors has the same state with the central element, reallocates the patch ID number and unify the patch ID number between the central element and the neighboring element when their states are the same.
(3) Repeat the second step until the total Patches ID number stabilized, which means any element has the same Patch ID number with their neighbors if they are both under presence state. Meanwhile, the area of each patch can be calculated based on the area of element. The results of patch analysis were shown in Table 6-5.

Table 6-5 Macroinvertebrate Patch Analysis

Patch Analysis	Discharge	
	28 m^3/s	60 m^3/s
Patches number	37	45
Max patch's area (m^2)	779359	137576
Min patch's area (m^2)	166	218
Average area (m^2)	30683	14135
Total areas (m^2)	1135276	636104

In this case, 'Four neighbors' scheme was applied in CA model which means that the four neighbors who have common sides with the central element are related cell of the patch. Of course, different neighboring scheme could be used if the patch classification criterion is changed.

6.3.2 Cellular automata Homogeneity

To further quantify the spatial distribution of *S. amurensis*, cellular automata homogeneity value (M.-Th. Hutt & R.Neff, 2001) was calculated in the study.

Due to the effects from local interaction in the CA model, we transformed the element's state of macroinvertebrate into a meta-state by certain rules. The corresponding rule is given with following equation: (Equation 6-5)

$$a_{ij} \rightarrow \frac{1}{|Nij|} \sum_{b \in N_{ij}} \theta(a_{ij}, b)$$

$$(6-5)$$

where a, b represent the state of central element a_{ij} and its nearest neighbors correspondingly, $|Nij|$ denotes the number of neighbors of central element a_{ij}.

In this case study, there is no distance difference in state-space, so the function θ became:

$$\theta(a,b) = \begin{cases} 1, & a = b \\ 0, & a \neq b \end{cases}$$

(6-6)

After transforming the macroinvertebrate's state in to a meta-state using CA technique, the CA homogeneity can be obtained for all the elements. Combining Equation (6.5) and Equation (6-6), and normalized summation among all elements, the CA homogeneity H was calculated by Equation (6-7):

$$H = \frac{1}{N}\sum_{ij}\frac{1}{|Nij|}\sum_{b \in N_{ij}}\theta(a_{ij},b)$$

(6-7)

By applying 'Four neighbors' scheme in the CA process, the homogeneity values for spatial distribution of macroinvertebrate are 0.8878 and 0.8907 for the discharge of 28 m³/s and 60 m³/s respectively. Meanwhile, the box homogeneity HB, which is related to the correlation length, was calculated, but the difference was not obvious. Due to the small patch number and some patches hold a big area, the box homogeneity HB is in fact meaningless in this research. In particular, the mesh size is already 20 m*20 m, there is little influence from far away neighbors.

6.4 Results and discussions

Results

This chapter simulated and analyzed the spatial distribution of macroinvertebrate under flow regulation. A hybrid model which integrates a two-dimensional water quality module with an ANN based habitat module was applied. Finally, the cellular automata technique was used to quantitatively analyze the spatial patterns of macroinvertebrate distribution.

The case study was at a compound reach in the middle of Lijiang River in China, where the discharge is regulated by the upstream Reservoir. The field data were collected during three years (2008-2010), and the model was used to analyze two typical flow regimes in dry season. The hybrid model efficiently characterized the distribution of S. amurensis under the two discharge conditions (28 m³/s and 60 m³/s). The results of hydro-environmental variables and macroinvertebrate distribution obtained from the model were in coincidence with the real observations.

The study showed there is a close relationship between environmental factors and macroinvertebrate distribution. Water depth, sediment and flow velocity determine the habitat conditions, and affect the composition and distribution of macroinvertebrates (Beauger et al, 2006). In this study, scenario analysis showed that the flow regulation had negative impacts on the spatial distribution of *Semisulcospira amurensis*, as the habitat area would decrease.

Discussion

This study maps macroinvertebrate presence with environmental factors based on an ANN modeling. There are some other alternatives such as statistics, fuzzy logic and genetic algorithm available to extract the non-linear relationships between species evolution and environmental factors. The effect of local interaction between nearest neighbors can be presented besides of the environmental factors in CA habitat evolution modeling. By coupling the numerical water quality modules with the habitat module, the hybrid model is able to investigate the influence of flow regulations on species habitats distribution.

Cellular automata technique was applied to characterize the spatial patterns of macroinvertebrate distribution. The patch analysis included the total patches number, patch area and so on. The results showed that when the discharge increased to 60 m^3/s, some small patches were formed, but the possible total presence area of S. amurensis decreased 44%. Homogeneity value was also calculated, but it was found not an appropriate index in this research due to the coarse spatial scale. The research results can give support to improve river management.

In this study, the discharge was constant in the two scenarios. However, the flow in reality is often a time series. In future, a hydrograph should be used to simulate the spatial and temporal dynamics of macroinvertebrate distribution. Moreover, the meshes should be refined to a reasonable scale.
The methodologies developed in this research can be used for macroinvertebrate as well as other aquatic organisms' distribution, if suitability data are available for these species.

Chapter 7

Individual-based and Spatial-based Unstructured Cellular Automata and application to aquatic ecosystem modelling

Traditional mathematical ecological models commonly rely on the well-mixed assumption or the mean-field (MF) approximation. It is well known that the MF assumption can simplify a complex system by replacing all interactions with the average interaction strength. While the MF assumption seems reasonable in some cases, it seems hardly accurate for other spatial heterogeneity situations (Li, 2009).

Besides, mean-field (MF) approximation will break down if the individuals by definition may only interact with each other within a limited neighborhood distance (E.g. forest trees). Many important features of ecological dynamics, such as the patterns of diversity and spatial distributions of species can be fundamentally changed when abandoning the MF assumption (Tilman and Kareiva, 1997). One way of relaxing the MF assumption is by formulating a spatially explicit individual-based model (IBM), or multi-agent systems, whose straightforward implementation is by means of a cellular automaton (Li et al., 2010).

Therefore, in order to relax the MF assumption and take into account spatial heterogeneity, this chapter considers the application of a spatially explicit individual-based model (IBM) as well as Unstructured Cellular Automata (UCA) to the same problem in aquatic ecology where the space introduces important information which simply cannot be neglected. (Lin et al., 2011b)

7.1 Description of study area

The case study was carried out with data form a small pond located in Deltares followed the study by Li (2009) and Li, et al. (2012). The size of the pond is about 52m*26m. In order to mimic the growth of water lilies in the small pond, time series of high resolution photos were recorded. The original photo is shown in Figure 7-1. The preliminarily process of photo used Image Processing Toolbox in Matlab environment. After photo feature was extracted, the photo became to Figure 7-2. Photo of week 18 (when the water lily began to appear in this year) was processed and taken as the initial condition for the modelling (see Figure7-4).

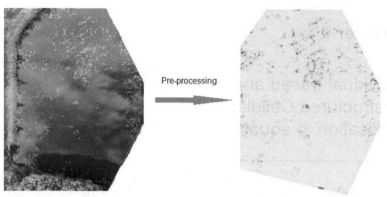

Fig. 7-1 Original photo   **Fig. 7-2** Photo feature Exacting

7.2 Influencing factors for water lily growth

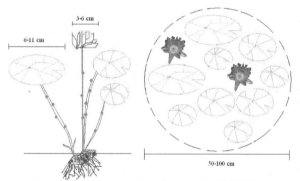

Fig. 7-3 spatial properties of water lily (Li, 2009)

In this study, the pond is quite small and it is assumed that there is no inflow and outflow, so the influence from wind and hydrodynamics was not taken into account. The main factors considered in this thesis which influence water lily growth are: (1) Water temperature (weekly averaged); (2) Sunshine duration (weekly accumulated); (3) Interaction between plant neighbors; (4) Physical properties of plants.

7.3 Spatial-based UCA Model setup

An Unstructured Cellular Automata (UCA) with 'three-sided' type scheme was used in the water lily model to capture growth patterns of water lily plants. The model was developed based on very fine unstructured meshes. In this case, 276392 unstructured meshes were established over the pond. The average size of grid element is 7cm by 7cm, which can represent the actual area of water lily leaves. The model time step is one week so that the simulation results from the modelling can be calibrated with the weekly high resolution photos.

The model was carried out for the year 2005 from week 18 when the water lily began to appear in this year. The spatial pattern of week 26 is shown in Figure 7-5. It is obvious that the water lily grew considerably and expanded during these weeks compared with the initial condition (Figure7-4). The snapshot of modelling at week 30 is illustrated in Figure 7-6, which can be compared with the real photo at the same week as shown in Figure 7-7.

Fig. 7-4 *Spatial pattern at week18 (Initial condition)*

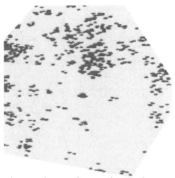

Fig. 7-5 Snapshot of model result at week26

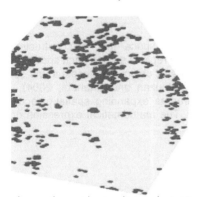

Fig. 7-6 Snapshot of model at week 30

Fig. 7-7 Photo at week 30

The result of the cellular automaton simulation model is seen to capture quite well the general pattern of the water lily spatially. Detailed experiments were carried out on, e.g. setting the rules for unstructured cellular automata modelling; whether the 'Moore' type neighbouring scheme of UCA should be employed instead of the 'three-sided' (Von Neumann) type scheme. At present, plant occupation analysis and plant pattern recognition procedures are being developed in the ongoing work.

Although unstructured cellular automata have great flexibility to present very complex boundaries, in this case study, the boundary is only a simple one and the advantages of UCA may not be obvious.

7.4 Individual Based Modelling using unstructured cellular automata

In nature, each individual has its own living-space. For plants, such as the water lily, one water lily can typically reach 1 meter height and cover a surface with a diameter ranging from 0.5 to 1.0 meter, with its flower having a diameter of 3 to 6 cm. The water lily has round and large leaves; one mature water lily leaf can have a diameter of about 6 to 11cm. Because of the immobility of water lilies, there might be a competition between two individuals for their living-space.

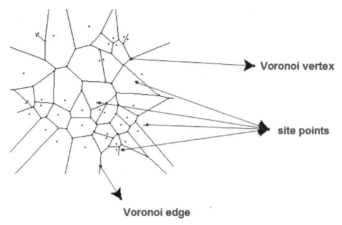

Voronoi vertex

site points

Voronoi edge

Fig. 7-8 Voronoi spatial model

The Voronoi spatial model is a tessellation of space that is constructed by decomposing the entire space into a set of Voronoi regions around each spatial object. By definition, points in the Voronoi region of a spatial object are closest to the spatial object than to any other spatial object (Kolahdouzan and Shahabi, 2004). The generations of Voronoi regions can be considered as 'expanding' spatial objects at a unique rate until these areas meet each other. The mathematical expression of the Voronoi region is defined as:

$$V(pi)=\{p \mid d(p,p_i) \le d(p,p_j), j \neq i, j=1...n\}$$

In this equation, the Voronoi region of spatial object pi, $V(p_i)$, is the region defined by the set of locations p in space where the distance from p to the spatial object pi, $d(p, p_i)$, is less than or equal to the distance from p to any other spatial object p_j.

In Fig. 7-9, the red crosses signify the initial positions of the plants and their Voronoi polygons (blue lines) represent the biggest living spaces if there is no competition among the neighbours. Actually the neighbours struggle for their living space as well; in the meantime their growth interferes with others. Assuming that there exists a big plant which occupies a large space, the birth of new plants can be restricted. When an old plant dies out, the point at which it lived is removed from the domain. On the other hand, when a new plant comes out, a new point is added to the domain. Therefore, birth and death rates are also important factors affecting the competition among plants in addition to their individual states and life-cycle.

From Fig. 7-10, it can be seen that plants are in different phases of their life-cycles, represented by different colours: "green" implies growing phase, "red" presents mature phase whereas "yellow" denotes the decay phase when gradually shrinks and finally vanishes.

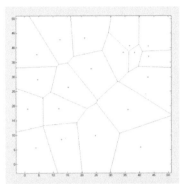

Fig. 7-9 Initial positions of water lily plants and their living space (2D view)

Fig. 7-10 Different phases of water lily in their life-cycles (2D view)

Fig. 7-11 Photo of week 19

Fig. 7-12 Snapshot of model at week 19

Fig. 7-13 Photo of week 26

Fig. 7-14 Snapshot of model at week 26

7.5 Analysis of results

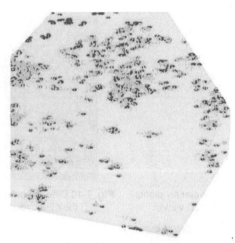

Fig. 7-15 Comparison between individual-based and spatial-based UCA model at week 26

By comparing Fig. 7-14 (individual-based model) with Fig. 7-5 (spatial-based model), which are snapshots at the same week (week 26) but come from different types of UCA modelling, different features can be observed from Fig 7-15. Since they considered the same influencing factors and evolution rules, both approaches can demonstrate and represent spatial patterns of water lily growths. But there are some distinctions between them. Fig. 7-14 (individual-based) shows every water lily plant in detail, while Fig. 7-5 (spatial-based) seems to easily cause bigger patchiness and loss individual features. During the simulation process when using these two models, the individual based model needed more running time when solving long-term problems, while the spatial-based UCA model cost some time for generating the mesh at first step but saving time for running model. In case detailed information is needed and the scale of issue is quite small, the Individual Based Modeling (IBM) approach is more suitable; conversely, the spatial based UCA modelling has the advantage for long term and large scale problems.

7.6 Future study

Despite UCA/CA had broad applications, the commercial software for CA in eco-hydraulics modelling is still ongoing, and most of them are two-dimensional (in the horizontal plane) and based on structured meshes.

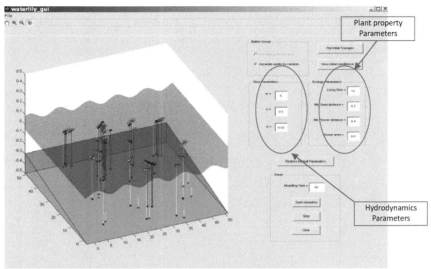

Fig. 7-16 Water lily modelling Graphic User Interface (3D view)

The above mentioned case study was with data from a small and shallow pond without much influence from the wind and dynamics of flow, therefore the vertical dimension is not needed. However, in many real cases, flow dynamics cannot be neglected in lakes therefore three dimensional modelling is needed.

Figure 7-16 shows a conceptual three-dimensional UCA model for water lily growing, in which the evolution of plants over the vertical dimension could be taken into growth account.

In this model, some stochastic factors (e.g. seed production and settlement) are considered, and some empirical values based on the physiological mechanisms of water lily are also introduced together with flow process, geometry factors etc.

The initial position of the water lilies could be obtained directly by extracting the features from real images. In order to test the sensitivity of the water lily model to initial conditions, random initial condition (with random special positions or different initial numbers of plants) was configured to run the unstructured cellular automata model. Hydrodynamic parameters and ecological parameters could be specified and adjusted in a graphical user interface of the water lily model (Figure 7-16), and their effects on UCA modelling could be observed from the evolutions of the water lily. The future studies will focus on a three dimensional model to include the influence of hydrodynamics in the UCA model.

Chapter 8

Conclusions and recommendations

Conclusions

Ecohydraulic modelling is about integrating hydrodynamic, water quality and eco-dynamic processes. These are affected by several factors and evolve spatially as well as temporally. Due to the complexity of these systems, it is hard to find general physical equations that describe all processes with sufficient detail. In the past few years, the Cellular Automata paradigm was applied to ecohydraulic systems and already proved to be a useful approach in ecohydraulics modelling (Minns et al., 2000; Mynett, 2002; Chen et al., 2004).

Most applications of cellular automata in ecohydraulic modelling are based on regularly spaced or structured computational grids. Recently, in computational modelling unstructured grids are become more popular due to their flexibility in representing complex geometries and arbitrary refinement. Moreover in reality, the computational area is usually irregular. This was the reason to explore unstructured cellular automata in this thesis.

In Chapter 2, the Classic Cellular Automata concepts with different structured lattices are presented. It was found that there is no obvious difference between structured triangular grids and structured square grids in a regular area, in the sense that both result in similar stable states. The influence of initial conditions vanishes rapidly within a few running steps.

Chapter 3 introduced the general concepts of Unstructured Cellular Automata and explored their behavior for a 'three cell-type' behaviour. The sensitivity of UCA was analyze and can be summarized as follows: (i) The dynamic competition reaches an equilibrium state, where the population of the three species roughly stabilizes around their average value; (ii) When the three species have similar initial populations, the ecosystem stabilizes very fast, with the populations oscillating remarkably around some fixed values (under-damped situation); (iii) When three species have different initial populations, the bigger the difference in their initial populations, the less the fluctuations are upon reaching the stable state. But in these cases, it takes longer time to reach the stable state (over-damped situation). In Section 3.5, different neighbourhood schemes of UCA which could also affect the patchiness pattern were depicted. Firstly, in terms of "Three-sided" type only considers the three neighbours and results in small patchiness appearances. However, in this case, all species can survive; i.e. none of them extinguish. No matter how non-uniform the initial condition is, the species keep competing all the time. In addition, because the "Three-sided" type is based on the small scale local rules, it usually leads to dynamic quasi-stable states. The principles inferred from numerous simulation studies can be summarized as follows. With the "Three-sided" rule, more stable population dynamics result with smaller patchiness, when compared with "Moore" type and "Three-Vertex" type. Meanwhile, when the more neighbours are taken into account, the more patchy patterns appear.

Usually, when developing models one needs to understand the basic underlying processes which is why partial differential equations are commonly used to model conservation processes. But in some cases, partial differential equations may have limitations. In contrast, complex nonlinear systems can be modelled by CA with a good computing efficiency. Chapter 4 gives some description of the computational theory of unstructured cellular automata, and a comparison is made between Partial Differential Equation and Unstructured Cellular Automata modelling. Compared with PDE-based models, Cellular Automata have the advantage that they can represent discrete entities directly and can reproduce emergent properties of behaviours with a large number of degrees of freedom. Another unique characteristic of CA-based modelling is that CA could generate dynamics patterns that are self-reproducing. It also should be mentioned that CA has universal parallel computing characteristics, which is powerful and has become an essential part in numerical computations.

The emphasis of Chapter 4 is on deducing the transition rules for cellular automata modelling from the PDE equivalent, using the finite difference method. Some types of partial differential equation were tested in this research. The result shows that under certain rules, the CA evolution results could be compared with the PDE performance. Taking into consideration that UCA has varying neighbours in contrast with classical CA, the influence of cell size in UCA was analysed in this thesis by the means of Finite Volume Method. The characteristic parameter —min distance of UCA— was put forward. Followed by several numerical experiments which were validated on different kinds of meshes (Rough meshes & local refined meshes & Global refined meshes), the characteristic parameter (min distance) for UCA proved to be useful to eliminate effects from varying neighbours and the simulation results turned out to be grid-independent.

Following up from the computational theory analysis of UCA, Chapter 5 gives three examples of using UCA in spatial dynamic ecological modelling, including (i) prey-predator modelling; (ii) algae bloom evolution modelling; and (iii) water quality modelling caused by spiked pollution loading in a bay area. Compared with traditional PDE based ecological modelling which usually focuses on mean-field (MF) approximations, these three applications illustrate the ability of UCA to represent processes with high spatio-temporal dynamics, which makes it especially suitable for processes with large spatial heterogeneity as well as for complex emergence.

Chapter 6 focuses on a practical case study of river restoration in China by quantifying the effect of flow regulation on the spatial distribution of macroinvertebrates. A hybrid model was set-up which integrates a two-dimensional water quality module with an ANN based habitat module. Later, the cellular automata technique was adopted to quantify the spatial distribution of the macroinvertebrates, more specifically, CA was used for calculating two criteria of spatial distribution. One is *patch analysis*, which includes the total number of patches, patch area and so on. The results show that after flow regulation, some small patches appeared while the total possible presence area of *S. amurensis* decreased by 44%. Another criterion for spatial distribution used in this research is *homogeneity*. This value represents the cluster characteristics of a full image (presence area together with absence area).

From this value, it can be seen that the increased discharge caused some scattered presence areas, while assembling (connecting) the absence areas of the habitat.

There are three types of numerical paradigms mainly used to represent spatial pattern dynamics (Ratze et al., 2007) namely (i) Partial Differential Equations (PDEs); (ii) Discrete paradigms including Cellular Automata (CA); and (iii) Agent-based paradigm including Individual-based models (IBMs). Among these three, the PDE and CA based paradigms were discussed in the first few chapters of this thesis. In order to compare UCA-based modelling with Individual-based modelling, Chapter 7 explored the applicability of UCA and IBM for an aquatic ecosystem case study of macrophyte growth in a small pond, using the same evolution rules. The results indicate that both approaches can adequately simulate and represent spatial pattern features of water lily growth. But there are some distinctions between them. IBM was able to capture every water lily plant in detail, while UCA modelling caused bigger patchiness and lost individual features. During the simulation process it was observed that the individual based model needed more running time when solving long-term problems, while the spatial-based UCA model cost some time for generating the mesh during the first steps, but saved considerable time when running model. In case detailed information is needed and the scale of issue is quite small, Individual Based Modelling (IBM) seems more suitable; conversely, the spatial based UCA modelling has the advantage for long-term and large-scale problems.

In summary, this thesis explored the Unstructured Cellular Automata paradigm, which is more flexible than classical cellular automata to better fit complicated geometries. This thesis illustrated several applications of UCA such as diffusion modelling, spatial ecological modelling and the hybrid eco-hydraulics modelling, all of them showing the ability of UCA to capture the spatial dynamic features quite well. Furthermore, this thesis reviewed the relationships between cellular automata with other types of modelling paradigms, including PDEs, IBMs and population growth modelling. Their advantages and disadvantages were listed and compared, and the specific characteristics of cellular automata were highlighted. In brief, this research revealed the capabilities of UCA and indicated that UCA can be useful as an alternative paradigm, especially for complex spatial dynamics problems.

Recommendations for future work

This thesis explored the relationships between PDE-based modelling and Unstructured Cellular Automata. Numerical experiments were carried out, but additional simulations using measured data needs to be carried out as well, to verify and compare the two approaches.

The spiked pollution problem mentioned in this thesis was simulated by UCA with rules extracted from the diffusion equation (PDE). The next step of this research could be to explore deducing the UCA rules considering the diffusion relations amongst neighbouring cells only. For example, the increasing/decreasing gradient could be used as a component of evolving UCA rules which could connect and determine the state values of the cell itself and its nearest neighbours. Perhaps in this way a more spiked diffusion pattern of effluents (like observed from field observations) could be captured, by dynamically adjusting the particular UCA rules.

A macroinvertebrate habitat model was developed in this research based on an artificial neural network (ANN) method (Chapter 6). In future work, the UCA meshes will be refined to a smaller scale requiring new CA rules for macroinvertebrate evolution modelling which combines the environmental factors with the local interaction between adjacent elements. By using cellular automata techniques in the habitat module, the interaction between neighbours of macroinvertebrate elements can indeed be taken into account, although It still needs attention to find the proper spatio-temporal dependence for extracting the proper evolution rules. In CA habitat modelling, not only hydro-environmental parameters affect the presence of macroinvertebrates, but also the influence from related neighbours. Moreover, the flow regulation value could be a time series so that the spatial distribution of macroinvertebrates will turn out to become a dynamic evolution process, which could perhaps be captured by a spatio-temporal hybrid model.

The case study of Water Lily growth occurred in a quite small and shallow square pond, so complex boundaries and geographical factors are not included in this case. In addition, within the small pond the influence from wind and hydrodynamics are very small so their effects were not taken into account in this thesis. For the broader application of UCA to such phenomena, it could be meaningful to develop a three-dimensional UCA modelling, in which vertical factors could also be taken into account.

In conclusion, the present research focused on exploring the concepts and feasibility of unstructured cellular automata. It was shown that UCA can extend the concept of cellular automata to irregular flow domains and arbitrary geometries. However, there is adequate room for more detailed studies, both on the mathematical concepts and on the practical applicability of unstructured cellular automata.

REFERENCES

Arashiro, E., Tome, T. (2007). The threshold of coexistence and critical behaviour of a predator–prey cellular automaton. *Journal of Physics A: Mathematical and Theoretical, 40*(5).

Aurenhammer, F., & Klein, R. (2000). Voronoi diagrams. In J. U. J.-R. Sack (Ed.), *Handbook of Computational Geometry* (pp. 201-290). Amsterdam Elsevier Science.

Babovic, V., Baretta, J. (1996). *Individual-based modelling of aquatic.* Paper presented at the International Conference on Hydroinformatics.

Balzter, H., Braun, P.W., Kohler, W. (1997). Cellular automata for vegetation dynamics. *Ecological Modelling, 107*, 113-125.

Bandman, Q. (2002). *Simulating spatial dynamics by probabilistic cellular automta.* Paper presented at the Theoretical and pratical issues on cellular automta: the 4th International Conference on Cellular automta for Research and Industry, London

Bar-Yam, Y. (1996). *Polymer simulation using Cellular Automata: 2-D Melts, Gel-Electrophoresis and Polymer collapse.* Paper presented at the Some New Directions in Science on Computers, Singapore.

Bartlett, M. S., & Hiorns, R. W. (1973). *Mathematical Theory of the Dynamics of Biological Populations*: Academic Press.

Bays, C. (1988). Classification of Semitotalistic Cellular Automata in Three Dimensions. *Complex Systems, 2.*

Beauger, A., Lair, N., Reyes-Marchant, P., & Peiry, J. L. (2006). The distribution of macroinvertebrate assemblages in a reach of the River Allier (France), in relation to riverbed characteristics. *Hydrobiologia, 571*(1), 63-76. doi: 10.1007/s10750-006-0217-x

Beigy, H., Meybodi, M. R. (2004). A Mathematical Framework for Cellular Learning Automata. *Advances in Complex Systems, 7*(3-4), 295-320.

Bennett, C., Grinstein, G. (1985). *Role of Irreversibility in Stabilizing Complex and Nonenergodic Behavior in Locally Interacting Discrete Systems.* Physical Review Letters.

Burks, A. W. (Ed.). (1970). *Von Neumann's self-reproduction automata.*

Carvalho, J. P., Carola, M., & Tomé, A. B. (2002). Forest Fire Modelling using Rule-Based Fuzzy Cognitive Maps and Voronoi Based Cellular Automata *FCT- Portuguese Foundation for Science and Technology.* Lisboa, Portugal

Cattaneo, G., Dennunzio, A., Farina, F. (2006). A Full Cellular Automaton to Simulate Predator-Prey Systems *Cellular Automata* (Vol. 4173, pp. 446-451): Springer Berlin Heidelberg.

Chen, Q., Yang, Q., Li, R., & Ma, J. (2013). Spring micro-distribution of macroinvertebrate in relation to hydro-environmental factors in the Lijiang River, China. *Journal of Hydro-environment Research, 7*(2), 103-112. doi: 10.1016/j.jher.2012.03.003

Chen, Q., Yang, Q., & Lin, Y. (2011). Development and application of a hybrid model to analyze spatial distribution of macroinvertebrates under flow regulation in the Lijiang River. *Ecological Informatics, 6*(6), 407-413. doi: 10.1016/j.ecoinf.2011.08.001

Chen, Q. W. (2004). *Cellular Automata and Artificial Intelligencein Ecohydraulics Modelling.* PhD, Delft University of Technology, the Netherlands.

Chen, Q. W. (2006). Stability Analysis of Harvesting Strategies in a Cellular Automata Based Predator-Prey Model.

Chen, Q. W. (2008). Cellular Automata *Handbook of Ecological Modelling and Informatics*: WIT Press.

Chen, Q. W., Mynett, A.E. (2003a). Integration of data mining techniques and heuristic knowledge in fuzzy logic modelling of eutrophication in Taihu Lake. *Ecological Modelling, 162* (1), 55-67.

Chen, Q. W., Mynett, A.E. (2004). *Modelling algal bloom in the dutch coast by integrated numerical and fuzzy crllular automata approaches.* Paper presented at the 6th Hydroinformatics Conference singapore.

Chen, Q. W., Mynett, A.E. (2006). Hydroinformatics Techniques in Eco-Environmental Modelling and Management *Journal of Hydroinformatics,, 8*(3), 297-316.

Chen, Q. W., Mynett. A.E. (2003b). Effects of cell size and configuration in cellular automata based prey–predator modelling. *Simulation Modelling Practice and Theory, 11*(7-8), 609–625.

Chen, Q. W., Mynett. A.E., Minns, A.W. (2002). *Coupling of scales in finite state cellular automata with applications to ecohydraulics modelling.* Paper presented at the Hydroinformatics Cardiff, UK.

Chon, T. S., Kwak, I.S., Song, M.Y., Park, Y.S., Cho, H.D, Kim, M.J., Cha, E.Y., Lek, S. (2002). Benthic macro invertebrates in streams of South Korea in different levels of pollution and patterning of communities by implementing the self-organizing mapping In D. Lee (Ed.), *Ecology of Korea* (pp. 356-384). Seoul: Bumwoo Publishing Company.

Chopard, B., Droz, M. (1998). *Cellular automata modelling of physical systems*: Cambridge University Press.

Chopard, B., Droz, M. . (1991). Cellular Automata Model for the Diffusion Equation. *Journal of Statistical Physics, Vol. 64*(Nos. 3/4), 859-892.

Codd, E. F. (1968). *Cellular Automata*: Academic Press.

Cortes, R.M.V., Ferreia, M.T., & Olivrira, S. V. (2002). Macroinvertebrate community structure in a regulated river segment with different flow conditions. *River Res. Applic., 18*, 367–382.

Dedecker, A. P., Goethals, P. L. M., Gabriels, W., & Pauw, N. D. (2004). Optimization of Artificial Neural Network (ANN) model design for prediction of macroinvertebrates in the Zwalm river basin (Flanders, Belgium). . *Ecological Modelling, 174,* 161-173.

Deutsch, E. S. (1972). Thinning Algorithms on Rectangular ? Hexagonal and Triangular Arrays. *Communication ACM, 15*(9), 827-837

Dunbar, M. J., Pedersen, M. L., Cadman, D., Extence, C., Waddingham, J., Chadd, R., & Larsen, S. E. (2010). River discharge and local-scale physical habitat influence macroinvertebrate LIFE scores. *Freshwater Biology Special Issue: ENVIRONMENTAL FLOWS: SCIENCE AND MANAGEMENT, 55*(1), 226-242.

Engrlen, G., White,R., Uljee,I. (1993). Integrating Constrained Cellular Automata Models, GIS and Decision Support Tools for Urban Planning and Policy Making. In H. Timmermans (Ed.), *Design and Decision Support Systims in Architecture and Urban Planning*. Dordrecht: Kluwer Academic

Ermentrout, G. B., Edelstein-Keshet, L. (1993). Cellular automata approaches to biological modeling. *J Theor Biol., 160*(1), 97-133.

Evans, M. R. (2012). Modelling ecological systems in a changing world. *Phil. Trans. Royal Society B, 367*, 181-190. doi: 10.1098/rstb.2011.0172

Farina, F., Dennunzio, A. (2008). A Predator-Prey Cellular Automaton with Parasitic Interactions and Environmental Effects. *Fundamenta Informaticae, 83*(4), 337-353.

Fathy, A. H. F., Aghababa, A.B. (2013). Cellular Learning Automata and Its Applications *Emerging Applications of Cellular Automata* (pp. 85-111).

Ferber, J. (1999). *Multi-Agent System: An Introduction to Distributed Artificial Intelligence.*

Fisch, R. (1992). Clustering in the One-Dimensional Three-Color Cyclic Cellular *Ann. Probab., 20*(3), 1528-1548.

Fischer, P. C. (1965). Generation of primes by a one-dimensional real-time iterative array. *Journal of the Association for Computing Machinery, 12*(3), 388-394.

Flache, A., & Hegselmann, R. (2001). Do Irregular Grids make a Difference? Relaxing the Spatial Regularity Assumption in Cellular Models of Social Dynamics. *Journal of Artificial Societies and Social Simulation 4*(4), 26.

Flocchini, P., Geurts, F., Mingarelli, A., Santoro, N. (2000). Covergence and aperioficity in fuzzy cellular automta : revisiting rule 90. *Physica D, 142*, 20-28.

Ganguly, N. (2003). *Cellular Automata Evolution: Theory and Applications in Pattern Recognition and Classification.* PhD, Bengal Engineering College Howrah, India.

Ganguly, N., Sikdar, B.K., Deutsch, A., Canright, G., Chaudhuri, P. (2003). *A Survey on Cellular Automata.*

Gardner, M. (1970). The fantastic combinations of John Conway's new solitaire game of life. *Scientific American, 220*(4), 120.

Green, D. G., Reichelt, R. E., Van der Laan, J., & Macdonald, B. W. (1989). *A generic approach to landscape modelling.* Paper presented at the 8th Biennial Conference of Simulation Society, Canberra.

Guinot, V. (2002). Modelling using stochastic, finite state cellular automata rule inference from continuum models. *Applied Mathematical Modelling 26*, 701-714.

Gutowitz, H. A. (Ed.). (1991). *Cellular Automtata: Theory and Experiment:* MIT Press.

Hütta, M.-T., & Neffb, R. (2001). Quantification of spatiotemporal phenomena by means of cellular automata techniques. *Physica A: Statistical Mechanics and its Applications, 289*(3-4), 498-516.

Ha, S., Ku, N., & Lee, K. Y. (2012). *Lattice Boltzmann Simulation for the Prediction of Oil Slick Movement and Spread in Ocean Environment.* Paper presented at the Twenty-second International Offshore and Polar Engineering Conference, Rhodes, Greece.

Hu, R. C., Ruan, X.G. (2003). *Differential Equation and Cellular Automata Model.* Paper presented at the International Conference on RoboticsJntelligent Systems and Signal Processing.

Itami, R. M. (1994). Simulating spatial dynamics: cellular automata theory *Landscape and Urban Planning, 30*, 27-47.

James, R. N. (1998). *Markov chains:* Cambridge University Press.

Kareafyllidis, I., Thanailakis, A. (1996). A model for predicting forest fire using cellular automata. *Ecological Modelling, 99*, 87-97.

Kingsbery, J. C. (2006). Excluded Blocks in Cellular Automata.

Kolahdouzan, M., & Shahabi, C. (2004). *Voronoi-Based K Nearest Neighbor Search for Spatial Network Databases.* Paper presented at the Proceedings of the 30th VLDB Conference, Toronto, Canada.

Langton, C. G. (1986). Studying Artificial Life with Cellular Automata. *Physica D, 22*.

Langton, C. G. (Ed.). (1995). *Artificial Life: An Overview:* MIT Press.

Lauren, M. K. (2001). Fractal Methods Applied to Describe Cellular Automaton Combat Models. *Fractals, 9*, 177-185.

Li, H. (2009). *Spatial Pattern Dynamics in Aquatic Ecosystem Modelling.* PhD, Delft University of Technology.

Li, H., Mynett, A. E., Qi, H., & Penning, E. (2010). Revealing spatial pattern dynamics in aquatic ecosystem modelling with multi-agent systems in Lake Veluwe. *Journal of Ecological Informatics, 5*(2), 97-107

Li, H., Mynett, A. E., & Ye, Q. (2012). Hydroinformatics in Multi-Colours --- Part Green: applications in aquatic ecosystem modelling. *Journal of Hydroinformatics,, 14*(4), 857–871.

Li, W., Han, R., Chen, Q., Qu, S., & Cheng, Z. (2010). Individual-based modelling of fish population dynamics in the river downstream under flow regulation. *Ecological Informatics, 5*(2), 115-123. doi: 10.1016/j.ecoinf.2009.12.006

Lin, Y., Mynett, A.E., Chen, Q. (2008). Application of Unstructured Cellular automata on ecological modelling. Paper presented at the 16th IAHR_APD Congress and 3rd Symposium of IAHR-ISHS, Nanjing, China.

Lin, Y., Mynett, A.E. (2010). *Performance of Different Neighbourhood Schemes in Unstructured Cellular Automata.* Paper presented at the 9th International Conference on Hydroinformatics, Tianjin, CHINA.

Lin, Y., Chen, Q., Yang, Q., Li, R. (2011a). *Macroinvertebrate spatial distribution under flow regulation using a hybrid model and its quantification by cellular automata techniques.* Paper presented at the Euromech 523 Ecohydraulics Colloquium (Ecohydraulics: linkages between hydraulics, morphodynamics and ecological processes in rivers), Clermont-Ferrand, France.

Lin, Y., Mynett, A.E., Li, H. (2011b). Unstructured Cellular Automata for modelling macrophyte Dynamics. *Journal of river basin management, 9*(3-4), 205-210.

Mandelbrot, B. (1967). How Long is the Coast of Britain? Statistical Self-Similarity and Fractional Dimension. *Science 156* (3775), 636-638. doi: 10.1126

McClain, M. E., Subalusky, A. L., Anderson, E. P., Dessu, S. B., Melesse, A. M., Ndomba, P. M., Mtamba, J. O. D., Tamatamah, R. A., Mligo, C. . (2014). Comparing flow regime, channel hydraulics and biological communities to infer flow-ecology relationships in the Mara River of Kenya and Tanzania. *Hydrological Sciences Journal.*

Meeker, L. E. (1998). *Four-dimensional Cellular Automata and the Game of Life*: University of South Carolina.

Miao, Z. (1997). The main environmental problems of Lijiang River. *Carsologica Sinica (In Chinese).*

Minns, A. W., Mynett, A.E., Chen, Q., Boogaard, H.F.P., van den. (2000). *A cellular automata approach to ecological modelling.* Paper presented at the Proceedings of Hydroinformarics Conference, Iowa, USA.

Mynett, A. E. (2002). *Environmental Hydroinformatics: The way ahead. (Keynote).* Paper presented at the the 5th International Conference on Hudroinformatics, Cardiff, UK.

Mynett, A. E., Chen, Q. (2004). *Integrated ecohydraulics modelling with application to marine eutrophication.* Paper presented at the The 5th International Ecohydraulics Conference, Madrid, Spain. .

Nagaya, T., Shiraishi, Y., Onitsuka, K., Higashino, M., Takami, T., Otsuka, N., Ozeki, H. (2008). Evaluation of suitable hydraulic conditions for spawning of ayu with horizontal 2D numerical simulation and PHABSIM. *Ecological Modelling, 215*, 134–143.

Ostrov, D. N., & Rucker, R. (1996). Continuous-valued cellular automata for nonlinear equations. *Complex Systems, 10*, 91-119.

Ostrovsky, B., Crooks, G. E., Smith, M. A., Bar-Yam Y. (2001). Cellular Automata for Polymer Simulation with Application to Polymer Melts and Polymer Collapse Including Implications for Protein Folding. *Parallel Computing, 27*, 613-641.

Perrier, J. Y., Sipper, M., Zahnd, J. . (1996). Toward a viable, self-reproducing iniversal computer. *Physica D, 97*, 335-352.

Postma, L., Stelling, G. S., & Boon, J. (1998). *3-Dimensional Water Quality and Hydrodynamic Modelling in Hong Kong III- Stratification and Water-Quality.* Paper presented at the 2nd International Symposium on Environmental Hydraulics, Hong Kong, China.

Ratze, C., Gillet, F., Muller, J.-P. and Stoffel, K. (2007). Simulation modelling of ecological hierarchies in constructive dynamical systems. *Ecological Complexity, 4*(1-2), 13-25.

Rorhman, D. H., Zeleski, S. (1997). *Latice-gas cellular automata: simple model of complex hydrodynamics*: Cambridge University Press.

Rosenfeld, A., & Wu, A. Y. (1980). *Reconfigurable Cellular Computers*: University of Maryland.

Rucker, R. (2003). Continuous-Valued Cellular Automata in Two Dimensions. In D. Griffeath, Moore, C. (Ed.), *New Constructions in Cellular Automata* Oxford University Press

Rušinović, Z., Bogunović, N. . (2006). *Cellular automata based model for the prediction of oil slicks behavior.*

Santucci, V. J., Gephard, S. R., & Pescitelli, S. M. (2005). Effects of Multiple Low-Head Dams on Fish, Macroinvertebrates, Habitat, and Water Quality in the Fox River, Illinois. *North American Journal of Fisheries Management 25:975–992, 2005, 25,* 975-992.

Seth, A., Bandyopadhyay, S., Maulik., U. . (2008). Probabilistic Analysis of Cellular Automata Rules and its Application in Pseudo Random pattern generation. *IAENG International Journal of Applied Mathematics, 38:4, IJAM_38_4_07.*

Sheldon, F., & Thoms, M. C. (2006). Relationships between flow variability and macroinvertebrate assemblage composition: data from four Australian dryland rivers. *River Research and Applications Special Issue: Variability in Riverine Environments, 22*(2), 219–238.

Shi, H. C., Shang, Y., & Chen, S. S. (2000). *A multi-agent system for computer science education.* Paper presented at the 5th annual SIGCSE/SIGCUE ITiCSEconference on Innovation and technology in computer science education, New York, NY, USA

Sieburg, H. B., McCutchan, J. A., Clay, K., Cabalerro, L., Osrlund, J. (1990). Simulation of HIV Infection in Artificial Immune System. *Physica D, 45,* 208-227.

Slimi, R., & Yacoubi, S. E. (2009). Spreadable cellular automata: modelling and simulations. *International Journal of Systems Science, 40*(5), 507-520.

Smith, A. R. (1969). Cellular Automata Theory *Technical Report No.2 Digital Systems Laboratory.* Procidence: Stanford University.

Stelling, G. S., 1984. . In: (Editor)., & Rijkswaterstaat, T. H. (1984). [On the construction of computational methods for shallow water flow problems].

Sternberg, S. R. (Ed.). (1980). *Language and Architecture for Parallel Image Processing.* North Holland.

Tilman, D., & Kareiva, P. (Eds.). (1997). *Spatial ecology : the role of space in population dynamics and interspecific interactions Monographs in population biology.* New Jersey: Princeton University Press.

Tobler, W. R. (1970). A computer movie simulating urban growth in the Detroit region. *Econ. Geogr., 46*(2), 234-240.

Toffoli, T., & Margolus, N. (1987). *Cellular Automata Machines: A new environment for modeling.*

Vichhac, G. (1984). Simulating Physics With Cellular Automata. *Physica D, 10,* 96-115.

Victor, J. D. (1990). What can automata theory tell us about the brain? *Physica D,* 205-227.

Volterra, V. (1926). Variazioni e fluttuazioni del numero d'individui in specie animali conviventi *Mem. Acad, 2,* 31–113.

Von Neumann, J. (1949). On Rings of Operators. Reduction Theory. *The Annals of Mathematics 2nd Ser. , 50*(2), 401-485. doi: 10.2307/1969463

Voss, R. F. (1984). Multiparticle fractal Aggregation *Journal of Statistical Physics, 36*(5/6), 861-872.

Wolfram, S. (1983). Statistical Mechanics of Cellular Automata. *Reviews of Modern Physics, 55*(3), 601-644.

Wolfram, S. (1984a). Computation theory of cellular automata. *Communication in Mathematical Physics, 96,* 15-57.

Wolfram, S. (1984b). Universality and complexity in cellular automata. *Physica D: Nonlinear Phenomena, 10*(1), 1-35.

Wolfram, S. (1985). Twenty Problems in the Theory of Cellular Automata. *Physica Scripta, Vol. T9,* 170-183.

Wolfram, S. (2002). *A new kind of Science.*

Wolfram, S. (Ed.). (1986). *Theory and Application of Cellular Automata.*

Wolfram, S. (Ed.). (1994). *Cellular Automata And Complexity: Collected Papers* Westview Press.

Wootton, J. T. (2001). Local Interactions predict large-scale patterns in empirically derived cellular automta. *Nature, 413,* 841-844.

Wortmann, J., Hearne, J.W., Adams, J.B. (1997). Evalutaion of effects of freshwater inflow on the distribution of estuarine macrophytes. *Ecological Modelling, 106,* 213-232.

Written, T. A., & Sander, L. M. (1983). Diffusion-Limited Aggregation *Physcal review B, 27*(9), 5686-5697.

Yang, X. S., & Young, Y. (2006). Cellular Automata,PDE, and Pattern Formation.pdf. In S. Olariu, Zomaya, A.Y. (Ed.), *Handbook of Bioinspired Algorithms and Applications* (pp. 271-281).

Ye, F., Chen, Q., & Li, R. (2010). Modelling the riparian vegetation evolution due to flow regulation of Lijiang River by unstructured cellular automata. *Ecological Informatics, 5*(2), 108-114. doi: 10.1016/j.ecoinf.2009.08.002

Appendices

Appendix A

Unstructured Cellular Automata based on triangle meshes

For Unstructured Cellular Automata based on triangular meshes, the experiments are designed with the conditions listed in the Table A-1, where the following design parameters are combined with different values: the initial rules (Three-sided, Moore, Three-vertex), the initial distributions (Randomly mixed, One isolated on boundary, One isolated in the middle), and the ratios of initial populations.

Table A-1 List of the simulation cases in Appendix-A.

Initial percentage	Initial distribution	Initial pattern	"Three-sided" rule	"Moore" rule	"Three-vertex" rule
Red 33% Green 33% White 34%	Randomly mixed	Fig. A-1	Fig. A-2	Fig. A-3	Fig. A-4
	One isolated on boundary	Fig. A-5	Fig. A-6	Fig. A-7	Fig. A-8
	One isolated in the middle	Fig. A-9	Fig. A-10	Fig. A-11	Fig. A-12
Red 20% Green 30% White 50%	Randomly mixed	Fig. A-13	Fig. A-14	Fig. A-15	Fig. A-16
	One isolated on boundary	Fig. A-17	Fig. A-18	Fig. A-19	Fig. A-20
	One isolated in the middle	Fig. A-21	Fig. A-22	Fig. A-23	Fig. A-24
Red 20% Green 40% White 40%	Randomly mixed	Fig. A-25	Fig. A-26	Fig. A-27	Fig. A-28
	One isolated on boundary	Fig. A-29	Fig. A-30	Fig. A-31	Fig. A-32
	One isolated in the middle	Fig. A-33	Fig. A-34	Fig. A-35	Fig. A-36
Red 20% Green 60% White 20%	Randomly mixed	Fig. A-37	Fig. A-38	Fig. A-39	Fig. A-40
	One isolated on boundary	Fig. A-41	Fig. A-42	Fig. A-43	Fig. A-44
	One isolated in the middle	Fig. A-45	Fig. A-46	Fig. A-47	Fig. A-48
Red 2% Green 49% White 49%	Randomly mixed	Fig. A-49	Fig. A-50	Fig. A-51	Fig. A-52
	One isolated on boundary	Fig. A-53	Fig. A-54	Fig. A-55	Fig. A-56
	One isolated in the middle	Fig. A-57	Fig. A-58	Fig. A-59	Fig. A-60

*The UCA experiments carried out cyclic rules with three colors, represented in section 3.2.1:
-----"Green" beat "White"; "White" beat "Red"; "Red" beat "Green".

A.1   The initial distribution by randomly mixed (Red33% Green33% White34%)

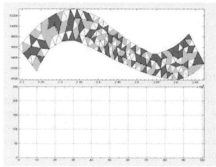

Fig. A-1 The initial distribution by randomly mixed (Red33% Green33% White34%)

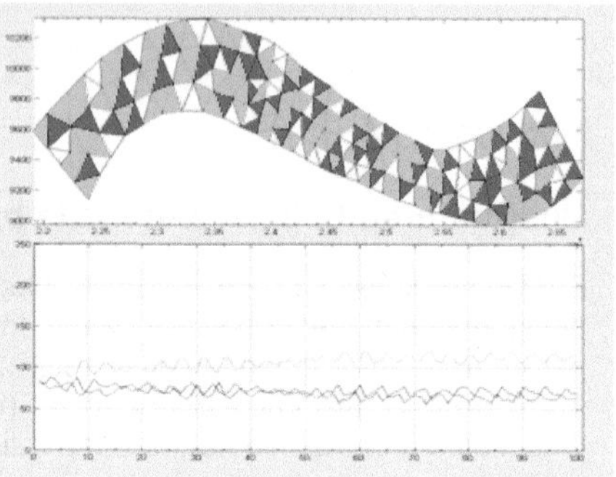

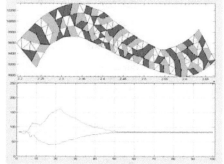

Fig. A-2 UCA implemented with 'Three-sided' rule

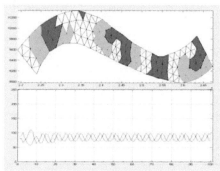

Fig. A-3 UCA implemented with 'Moore' rule

Fig. A-4 UCA implemented with 'Three-vertices' rule

Initial percentage	Initial distribution	Initial pattern	"Three-sided" rule	"Moore" rule	"Three-vertex" rule
Red 33% Green 33% White 34%	*Randomly mixed*	*Fig. A-1*	Fig. A-2	Fig. A-3	Fig. A-4
Spatial Distribution			no patchiness	obvious patchiness	likely patchiness
Population Dynamic			fluctuate around some seemingly stable values	stabilized quite quickly	stabilizes around their average

A.2 The initial distribution with one isolated on the boundary based on triangle meshes (Red33% Green33% White34%)

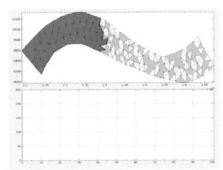

Fig. A-5 The initial distribution with one isolated on the boundary

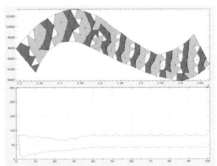

Fig. A-6 UCA implemented with 'Three-sided' rule

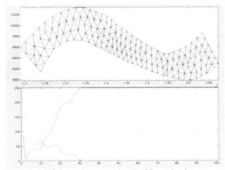

Fig. A-7 UCA implemented with 'Moore' rule

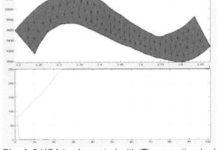

Fig. A-8 UCA implemented with 'Three-vertices' rule

Initial percentage	Initial distribution	Initial pattern	**"Three-sided" rule**	**"Moore" rule**	**"Three-vertex" rule**
Red 33% Green 33% White 34%	*One isolated on boundary*	*Fig. A-5*	Fig. A-6	Fig. A-7	Fig. A-8
Spatial Distribution			seemingly patchiness	one domination (white)	one domination (red)
Population Dynamic			stable	two extinctions	two extinctions

A.3 The initial distribution with one isolated in the middle based on triangle meshes (Red33% Green33% White34%)

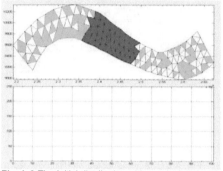

Fig. A-9 The initial distribution with one isolated on in the middle

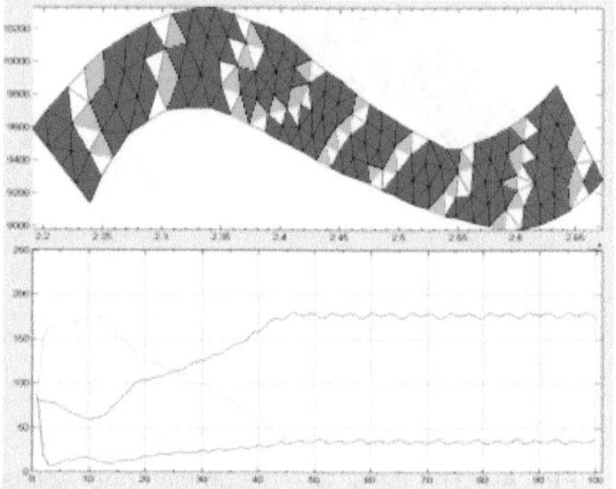

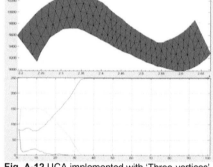

Fig. A-10 UCA implemented with 'Three-sided' rule

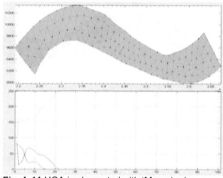

Fig. A-11 UCA implemented with 'Moore' rule

Fig. A-12 UCA implemented with 'Three-vertices' rule

Initial percentage	Initial distribution	Initial pattern	"Three-sided" rule	"Moore" rule	"Three-vertex" rule
Red 33% Green 33% White 34%	One isolated in the middle	Fig. A-9	Fig. A-10	Fig. A-11	Fig. A-12
Spatial Distribution			bigger patchiness (red)	one domination (green)	one domination (red)
Population Dynamic			Stabilized	two extinctions	two extinctions

A.4 The initial distribution by randomly mixed based on triangle meshes (Red20% Green30% White50%)

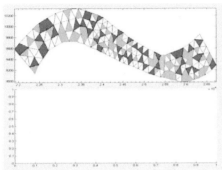

Fig. A-13 The initial distribution by randomly mixed

Fig. A-14 UCA implemented with 'Three-sided' rule

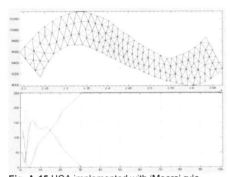

Fig. A-15 UCA implemented with 'Moore' rule

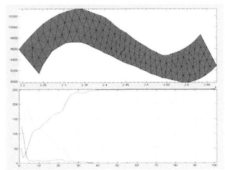

Fig. A-16 UCA implemented with 'Three-vertices' rule

Initial percentage	Initial distribution	Initial pattern	"Three-sided" rule	"Moore" rule	"Three-vertex" rule
Red 20% Green 30% White 50%	Randomly mixed	Fig. A-13	Fig. A-14	Fig. A-15	Fig. A-16
Spatial Distribution			no patchiness	one domination (white)	one domination (red)
Population Dynamic			convergence to the average value but cost long time	two extinctions	two extinctions

A.5 The initial distribution with one isolated on the boundary based on triangle meshes (Red20% Green30% White50%)

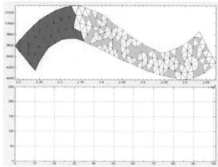

Fig. A-17 The initial distribution with one isolated on the boundary

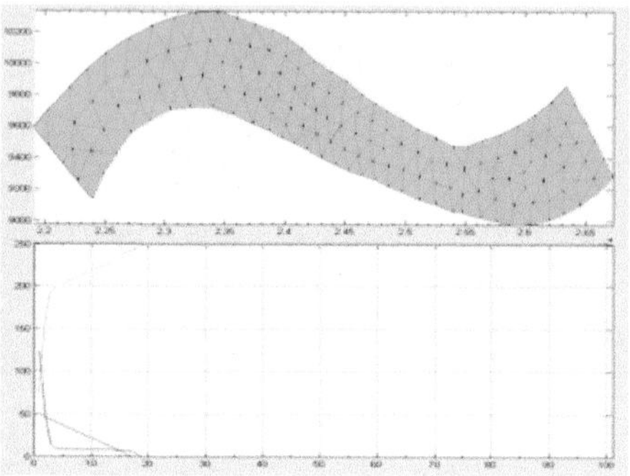

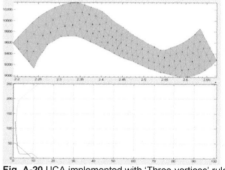

Fig. A-18 UCA implemented with 'Three-sided' rule

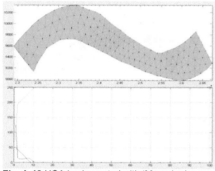

Fig. A-19 UCA implemented with 'Moore' rule

Fig. A-20 UCA implemented with 'Three-vertices' rule

Initial percentage	Initial distribution	Initial pattern	"Three-sided" rule	"Moore" rule	"Three-vertex" rule
Red 20% Green 30% White 50%	One isolated on boundary	Fig. A-17	Fig. A-18	Fig. A-19	Fig. A-20
Spatial Distribution			one domination (green)	one domination (green)	one domination (green)
Population Dynamic			two extinctions	two extinctions	two extinctions

A.6   The initial distribution with one isolated in the middle based on triangle meshes (Red20% Green30% White50%)

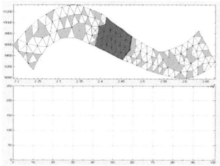

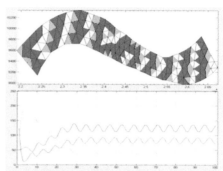

Fig. A-21 The initial distribution with one isolated on in the middle

Fig. A-22 UCA implemented with 'Three-sided' rule

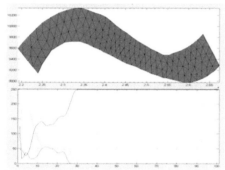

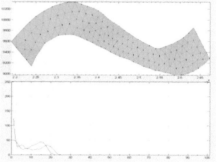

Fig. A-23 UCA implemented with 'Moore' rule

Fig. A-24 UCA implemented with 'Three-vertices' rule

Initial percentage	Initial distribution	Initial pattern	"Three-sided" rule	"Moore" rule	"Three-vertex" rule
Red 20% *Green 30%* *White 50%*	*One isolated in the middle*	*Fig. A-21*	Fig. A-22	Fig. A-23	Fig. A-24
Spatial Distribution			close to patches	one domination (red)	one domination (green)
Population Dynamic			stable with separate value	two extinctions	two extinctions

A.7 The initial distribution by randomly mixed based on triangle meshes (Red 20% Green40% White40%)

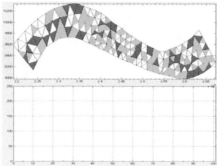

Fig. A-25 The initial distribution by randomly mixed

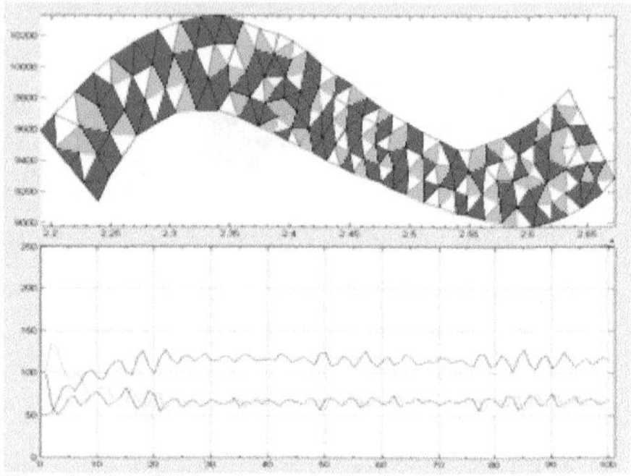

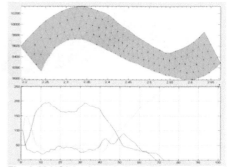

Fig. A-26 UCA implemented with 'Three-sided' rule

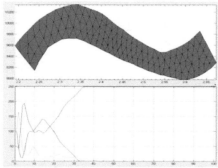

Fig. A-27 UCA implemented with 'Moore' rule

Fig. A-28 UCA implemented with 'Three-vertices' rule

Initial percentage	Initial distribution	Initial pattern	"Three-sided" rule	"Moore" rule	"Three-vertex" rule
Red 20% Green 40% White 40%	*Randomly mixed*	*Fig. A-25*	Fig. A-26	Fig. A-27	Fig. A-28
Spatial Distribution			no obvious patchiness	one domination (red)	one domination (green)
Population Dynamic			dynamic stable	two extinctions	two extinctions after some competitions

A.8 The initial distribution with one isolated on the boundary based on triangle meshes (Red 20% Green40% White40%)

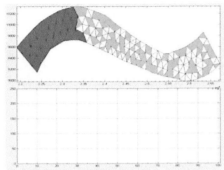

Fig. A-29 The initial distribution with one isolated on the boundary

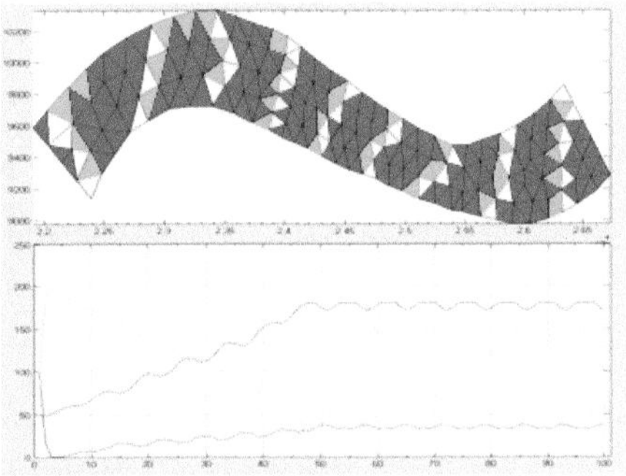

Fig. A-30 UCA implemented with 'Three-sided' rule

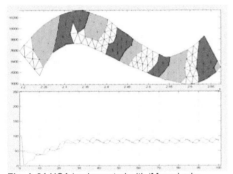

Fig. A-31 UCA implemented with 'Moore' rule

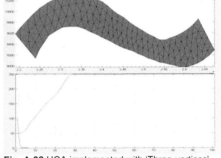

Fig. A-32 UCA implemented with 'Three-vertices' rule

Initial percentage	Initial distribution	Initial pattern	"Three-sided" rule	"Moore" rule	"Three-vertex" rule
Red 20% Green 40% White 40%	One isolated on boundary	Fig. A-29	Fig. A-30	Fig. A-31	Fig. A-32
Spatial Distribution			red cause bigger patchiness	fairly patchiness	one domination (red)
Population Dynamic			stable with higher red population	convergence around average	two extinctions (white and green)

A.9 The initial distribution with one isolated in the middle based on triangle meshes (Red 20% Green40% White40%)

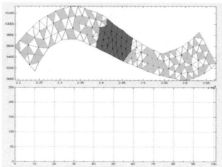

Fig. A-33 The initial distribution with one isolated on in the middle

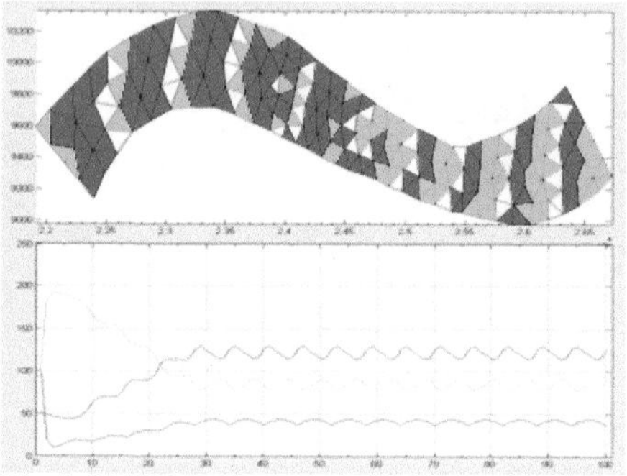

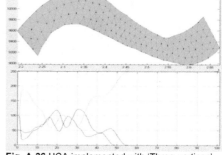

Fig. A-34 UCA implemented with 'Three-sided' rule

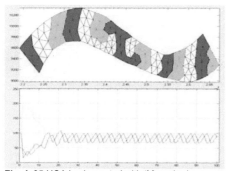

Fig. A-35 UCA implemented with 'Moore' rule

Fig. A-36 UCA implemented with 'Three-vertices' rule

Initial percentage	Initial distribution	Initial pattern	"Three-sided" rule	"Moore" rule	"Three-vertex" rule
Red 20% Green 40% White 40%	One isolated in the middle	Fig. A-33	Fig. A-34	Fig. A-35	Fig. A-36
Spatial Distribution			likely patchiness	patchiness	one domination (green)
Population Dynamic			stable with respective population	fluctuate around stable values	two extinctions (white and red)

A.10 The initial distribution by randomly mixed based on triangle meshes (Red 20% Green60% White20%)

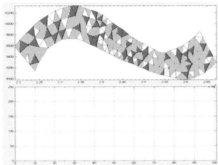

Fig. A-37 The initial distribution by randomly mixed

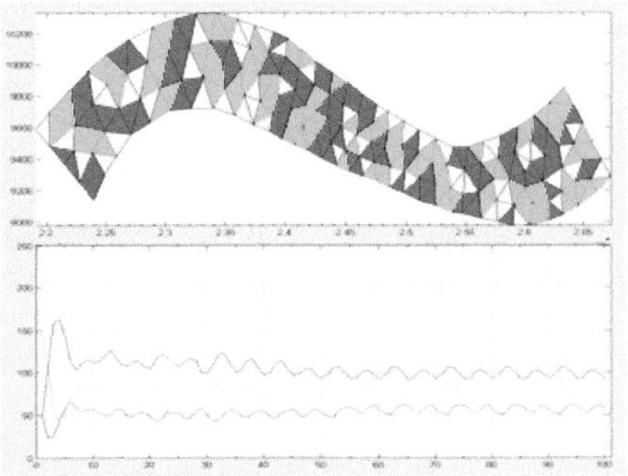

Fig. A-38 UCA implemented with 'Three-sided' rule

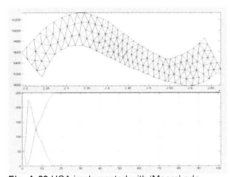

Fig. A-39 UCA implemented with 'Moore' rule

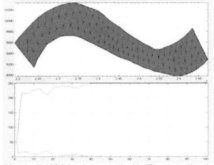

Fig. A-40 UCA implemented with 'Three-vertices' rule

Initial percentage	Initial distribution	Initial pattern	"Three-sided" rule	"Moore" rule	"Three-vertex" rule
Red 20% Green 60% White 20%	Randomly mixed	Fig. A-37	Fig. A-38	Fig. A-39	Fig. A-40
Spatial Distribution			No patchiness	one domination (white)	one domination (red)
Population Dynamic			Seemingly stable	two extinctions	two extinctions

A.11 The initial distribution with one isolated on the boundary based on triangle meshes (Red 20% Green60% White20%)

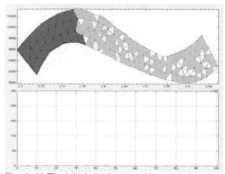

Fig. A-41 The initial distribution with one isolated on the boundary

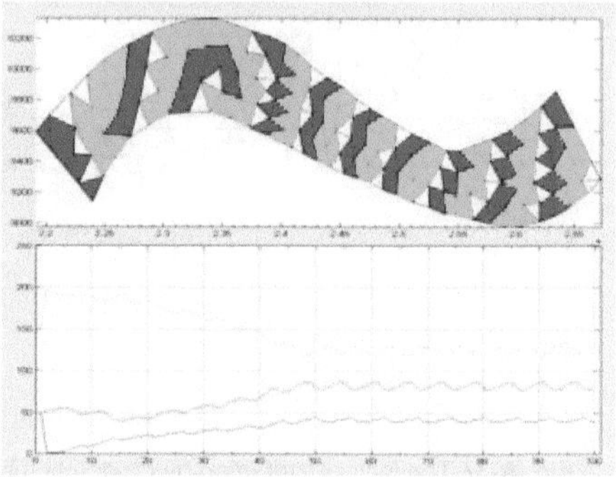

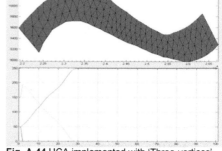

Fig. A-42 UCA implemented with 'Three-sided' rule

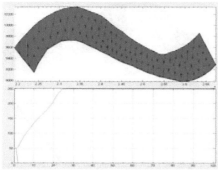

Fig. A-43 UCA implemented with 'Moore' rule

Fig. A-44 UCA implemented with 'Three-vertices' rule

Initial percentage	Initial distribution	Initial pattern	"Three-sided" rule	"Moore" rule	"Three-vertex" rule
Red 20% Green 60% White 20%	One isolated on boundary	Fig. A-41	Fig. A-42	Fig. A-43	Fig. A-44
Spatial Distribution			seemingly patchiness	one domination (red)	one domination (red)
Population Dynamic			stable with respective population	two extinctions	two extinctions

A.12   The initial distribution with one isolated in the middle based on triangle meshes
(Red 20% Green60% White20%)

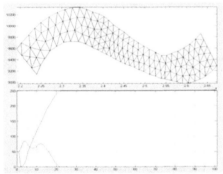
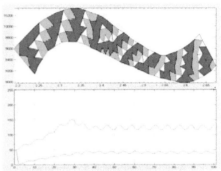

Fig. A-45 The initial distribution with one isolated on in the middle

Fig. A-46 UCA implemented with 'Three-sided' rule

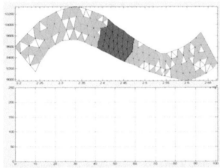
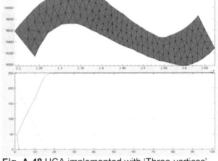

Fig. A-47 UCA implemented with 'Moore' rule

Fig. A-48 UCA implemented with 'Three-vertices' rule

Initial percentage	Initial distribution	Initial pattern	"Three-sided" rule	"Moore" rule	"Three-vertex" rule
Red 20% Green 60% White 20%	*One isolated in the middle*	*Fig. A-45*	Fig. A-46	Fig. A-47	Fig. A-48
Spatial Distribution			seemingly patchiness	one domination (white)	one domination (red)
Population Dynamic			stable with respective population	two extinctions	two extinctions

A.13 The initial distribution by randomly mixed based on triangle meshes (Red 2% Green49% White49%)

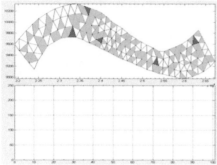

Fig. A-49 The initial distribution by randomly mixed

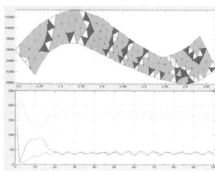

Fig. A-50 UCA implemented with 'Three-sided' rule

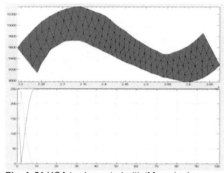

Fig. A-51 UCA implemented with 'Moore' rule

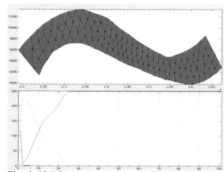

Fig. A-52 UCA implemented with 'Three-vertices' rule

Initial percentage	Initial distribution	Initial pattern	"Three-sided" rule	"Moore" rule	"Three-vertex" rule
Red 2% Green 49% White 49%	Randomly mixed	Fig. A-49	Fig. A-50	Fig. A-51	Fig. A-52
Spatial Distribution			Green cause bigger patchiness	one domination (red)	one domination (red)
Population Dynamic			stable with higher green population	two extinctions (white and green)	two extinctions (white and green)

A.14 The initial distribution with one isolated on the boundary based on triangle meshes (Red 2% Green49% White49%)

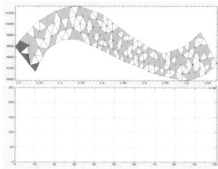

Fig. **A-53** The initial distribution with one isolated on the boundary

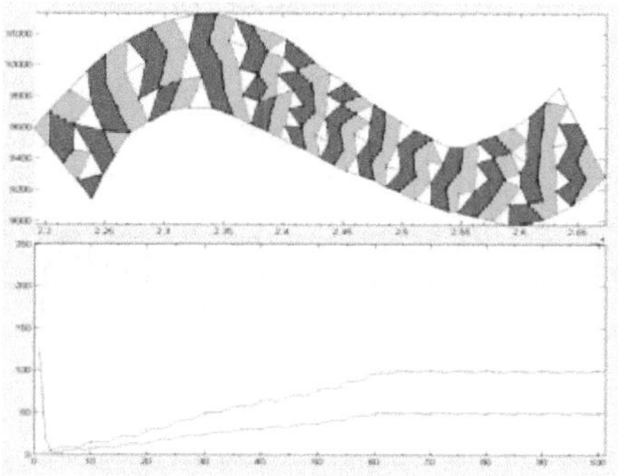

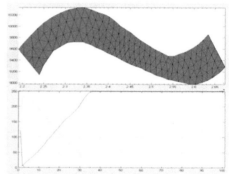

Fig. **A-54** UCA implemented with 'Three-sided' rule

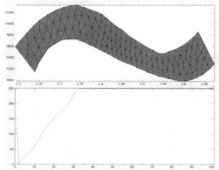

Fig. **A-55** UCA implemented with 'Moore' rule

Fig. **A-56** UCA implemented with 'Three-vertices' rule

Initial percentage	Initial distribution	Initial pattern	"Three-sided" rule	"Moore" rule	"Three-vertex" rule
Red 2% Green 49% White 49%	One isolated on boundary	Fig. A-53	Fig. A-54	Fig. A-55	Fig. A-56
Spatial Distribution			dynamic patchiness	one domination (red)	one domination (red)
Population Dynamic			stable	two extinctions	two extinctions

A.15 The initial distribution with one isolated in the middle based on triangle meshes (Red 2% Green49% White49%)

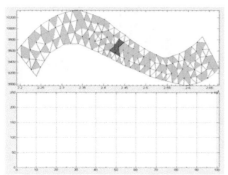

Fig. A-57 The initial distribution with one isolated on in the middle

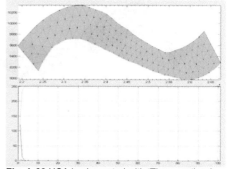

Fig. A-58 UCA implemented with 'Three-sided' rule

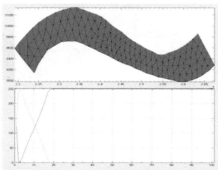

Fig. A-59 UCA implemented with 'Moore' rule

Fig. A-60 UCA implemented with 'Three-vertices' rule

Initial percentage	Initial distribution	Initial pattern	"Three-sided" rule	"Moore" rule	"Three-vertex" rule
Red 2% Green 49% White 49%	One isolated in the middle	Fig. A-57	Fig. A-58	Fig. A-59	Fig. A-60
Spatial Distribution			Green cause bigger patchiness	one domination (red)	one domination (green)
Population Dynamic			Stable with higher green population	two extinctions	two extinctions

Appendix   B

Unstructured Cellular Automata based on polygon meshes

For Unstructured Cellular Automata which based on polygon elements, the experiments are designed with the following design parameters: the initial distributions (Randomly mixed, One isolated on boundary, One isolated in the middle), and the ratios of initial populations.

Initial percentage	Initial distribution	Initial pattern	UCA based on polygon elements
Red 20% Green 30% White 50%	Randomly mixed	Fig. B-1	Fig. B-2
	One isolated on boundary	Fig. B-3	Fig. B-4
	One isolated in the middle	Fig. B-5	Fig. B-6
Red 20% Green 60% White 20%	Randomly mixed	Fig. B-7	Fig. B-8
	One isolated on boundary	Fig. B-9	Fig. B-10
	One isolated in the middle	Fig. B-11	Fig. B-12

The polygon element based Unstructured Cellular Automata has the advantage of balancing the equilibrium and manifests the patchiness characteristics. It also takes more neighbors into consideration. Compared with triangle-based unstructured cellular automata, this kind of paradigm shows bigger patchiness and it's more likely become stable.

B.1 The initial distribution based on polygon elements with percentage:
Red20% Green30% White50%

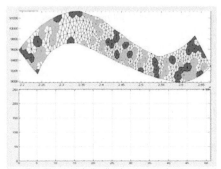

Fig. B-1 The initial distribution by randomly mixed based on polygon elements

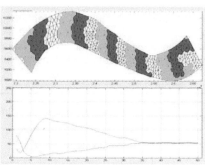

Fig. B-2 UCA implemented with initial distribution by randomly mixed

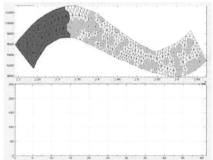

Fig. B-3 The initial distribution with one isolated on the boundary

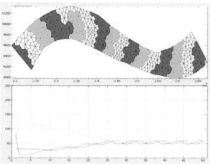

Fig. B-4 UCA implemented with initial distribution by one isolated on the boundary

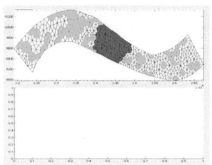

Fig. B-5 The initial distribution with one isolated in the middle

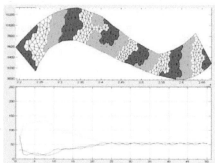

Fig. B-6 UCA implemented with initial distribution by one isolated in the middle

B.2 The initial distribution based on polygon elements with percentage:
Red20% Green60% White20%

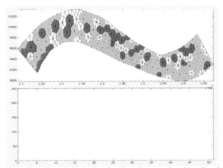

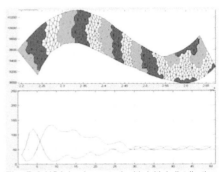

Fig. B-7 The initial distribution by randomly mixed based on polygon elements

Fig. B-8 UCA implemented with initial distribution by randomly mixed

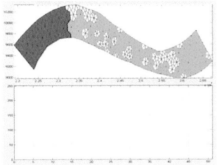

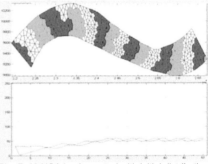

Fig. B-9 The initial distribution with one isolated on the boundary based on polygon elements

Fig. B-10 UCA implemented with initial distribution by one isolated on the boundary)

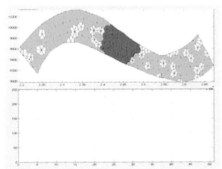

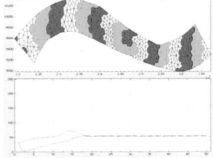

Fig. B-11The initial distribution with one isolated in the middle based on polygon elements

Fig. B-12 UCA implemented with initial distribution by one isolated in the middle

About the Author

Yuqing Lin was born on 18 September 1982 in the beautiful and historic city Yangzhou in Jiangsu Province, China. She obtained her bachelor's degree in Hydrology and Water Resources from Hohai University in Nanjing. Because of her good achievements she was recommended to become a graduate student without examination, in 2004. During her graduate studies at Hohai, her research interests were in hydroinformatics and hydrodynamic modelling.

In 2007 she obtained a fellowship from the China Scholarship Council for studying in the Netherlands, and she obtained her Master of Science degree in Hydroinformatics (with distinction) under the supervision of and Prof. Mynett in 2008 at UNESCO-IHE. She continued as a PhD fellow with financial support from Deltares, Chinese Academy of Science and UNESCO-IHE. She encountered some health problems during 2011-2012, but recovered in 2013 and continued her research on 'Unstructured Cellular Automata in Ecohydraulics Modelling'. Her research interests include: mathematical modelling, ecohydraulics, and environmental hydroinformatics.

Publications

Li, G., Lin Y. (2005). *A hydrodynamic method used in river basin flood-routing modelling.* Paper presented at the Second International Symposium on Methodology in Hydrology Nanjing, China.

Lin, Y., Mynett, A.E., Chen, Q. (2008). *Application of unstructured cellular automata to ecological modelling.* Paper presented at the 26th IAHR–APD Congress and 3rd Symposium of IAHR–ISHS, Nanjing, China.

Mynett, A. E., Lin, Y., Chen Q. (2009). *Unstructured cellular automata for enhanced population dynamics modelling.* Paper presented at the 7th International Symposium on Ecohydraulics & 8th International Conference on Hydroinformatics, Concepción, Chile.

Lin, Y., Mynett, A.E. (2010a). *Analysis of cell size influence in unstructured cellular automata modelling.* Paper presented at the 9th International Conference on Hydroinformatics, Tianjin,China.

Lin, Y., Mynett, A.E. (2010b). *Effects of Cell Size in unstructured cellular automata.* Paper presented at the First IAHR European Congress, Edinburgh, UK.

Lin, Y., Mynett, A.E. (2010c). *Performance of Different Neighbourhood Schemes in Unstructured Cellular Automata.* Paper presented at the 9th International Conference on Hydroinformatics, Tianjin, CHINA.

Lin, Y., Mynett, A.E. (2010d). *Spatial water quality modelling using unstructured cellular automata for spiked pollution loading in Hong Kong.* Paper presented at the 6th International Symposium on Environmental Hydraulics, Greece.

Lin, Y., Mynett, A.E., Li, H. (2010). *Ecohydraulic Modelling Based On Unstructured Cellular Automata.* Paper presented at the 8th International Symposium on Ecohydraulics, Seoul, Korea.

Mynett, A. E., Lin, Y. (2010). *Adaptive Meshing in Unstructured Cellular Automata.* Paper presented at the 9th International Conference on Hydroinformatics, Tianjin, China.

Lin, Y., Chen, Q., Yang, Q., Li, R. (2011). *Macroinvertebrate spatial distribution under flow regulation using a hybrid model and its quantification by cellular automata techniques.* Paper presented at the Euromech 523 Ecohydraulics Colloquium, Clermont-Ferrand, France.

Lin, Y., Mynett, A.E., Li, H. (2011). Unstructured Cellular Automata for modelling macrophyte Dynamics. *Journal of River Basin Management, 9*(3-4), 205-220.

Chen, Q., Yang, Q., Lin, Y. (2011). Development and application of a hybrid model to analyze spatial distribution of macroinvertebrates under flow regulation in the Lijiang River. *Ecological Informatics, 6*, 407-423.

Lin, Y., Chen, Q., Yang, Q. (2014a). Quantifying spatial distributions of macroinvertebrate under flow regulation by unstructured cellular automata (in preparation).

Lin, Y., Mynett, A.E. (2014b). A study on the computational theory of unstructured cellular automata (in preparation).

Lin, Y., Mynett, A.E., Chen, Q. (2014c). A comparison of cellular automata with partial differential equation based modelling (in preparation).

T - #0436 - 101024 - C128 - 240/170/7 - PB - 9781138027404 - Gloss Lamination